Artificial Intelligence: A Guide for Everyone

W0111419

Arshad Khan

Artificial Intelligence: A Guide for Everyone

 Springer

Arshad Khan
University of California
San Diego, CA, USA

ISBN 978-3-031-56712-4   ISBN 978-3-031-56713-1   (eBook)
https://doi.org/10.1007/978-3-031-56713-1

This Springer imprint is published by the registered company Springer Nature Switzerland AG
The registered company address is: Gewerbestrasse 11, 6330 Cham, Switzerland

If disposing of this product, please recycle the paper.

Preface

For decades, there have been many attempts to understand artificial intelligence (AI). The progress has been slow and measurable. However, since the release of ChatGPT in 2022, the deployment and use of AI have skyrocketed. Its emergence as a potent force is challenging the boundaries of what is possible.

Enterprises, as well as individuals, are racing to reap the benefits of AI. However, in most cases, they are doing so without understanding the technology or its implications and risks, which can be significant. *Artificial Intelligence: A Guide for Everyone* is a step in addressing that gap by providing information that readers can easily understand at every level.

This book aims to provide useful information to those planning, developing, or using AI, which has the potential to transform industries and shape the future. Whether you are stepping into the world of AI for the first time or are a seasoned professional seeking deeper insights, this comprehensive guide ensures that both beginners and experienced individuals find value within its pages.

Artificial Intelligence: A Guide for Everyone encompasses theoretical as well as practical aspects of AI across various industries and applications. It demystifies AI by explaining, in a language that non-techies can follow, its different types, differentiating technologies, and various aspects of implementation. It explains the connection between AI theory and real-world application across diverse industries and how it fuels innovation.

Whether you are a simple business user, a professional, a tech enthusiast, a manager, an executive, or an aspiring AI developer, this guide is your roadmap to navigate the intricacies of AI. The topics covered in this book are comprehensive and diverse. They include its history, benefits, and disadvantages, as well as its relationship and impact on humans.

This guide explains the prerequisites for building, developing, and deploying an AI system. It discusses its key techniques, technologies, and components, such as machine learning, natural language processing, and deep learning. It explains the process of constructing an AI system, offering insights into problem definition, data preprocessing, model development, evaluation, testing, deployment, and monitoring. In consideration of practical resource limitations and constraints, pre-built, ready-to-use AI solutions, which democratize AI, are discussed.

This book also delves into the intricacies of measuring AI, exploring various methods and metrics to assess human likeness, task performance, and ethical considerations. The exploration extends to the comparison of measurement methods and the diverse approaches used to evaluate AI performance.

The journey concludes by categorizing AI based on intelligence levels and functionality, providing a comprehensive understanding of Artificial Narrow Intelligence (ANI), Artificial General Intelligence (AGI), Artificial Super Intelligence (ASI), and other categories.

Artificial Intelligence: A Guide for Everyone serves as a compass through the ever-evolving landscape of artificial intelligence. This guide will help illuminate the multifaceted aspects of this transformative technology for any novice exploring the world of AI as well as for a seasoned professional navigating its complexities.

San Diego, CA, USA   Arshad Khan

Contents

1

Introduction

Artificial Intelligence

Definition

Artificial intelligence (AI) is a term coined in 1955 by John McCarthy, a Stanford professor. He defined it as "The science and engineering of making intelligent machines, especially intelligent computer programs." Over the years, many other definitions have been proposed, including the following:

- Artificial intelligence is intelligence demonstrated by machines, as opposed to the natural intelligence displayed by animals and humans.
- Artificial intelligence means a machine-based system that can, for a given set of human-defined objectives, make predictions, recommendations, or decisions influencing real or virtual environments.
- Artificial intelligence involves the theory and development of computer systems able to perform tasks that normally require human intelligence, such as visual perception, speech recognition, decision-making, and translation between languages.

Approaches

Russell and Norvig, authors of *Artificial Intelligence: A Modern Approach*, have identified four potential goals or definitions of AI, which differentiate computer systems based on rationality and thinking versus acting:

- Human approach:

  - Systems that think like humans.
  - Systems that act like humans.

- Ideal approach:

  - Systems that think rationally.
  - Systems that act rationally.

Objective

The objective of AI is to create technology that allows computers and machines to work intelligently so that they can perform tasks that usually require human intelligence. It aims to enhance human capabilities and perform intellectual tasks like providing solutions, answering questions, making predictions, solving problems, decision-making, offering strategic suggestions, and understanding human communication. It aims to understand languages, recognize patterns, make decisions, solve problems, and learn from experience. AI's goal is to make machines capable of mimicking and even enhancing human cognitive abilities, ultimately making our lives easier, more efficient, and more productive.

AI Simplified

Artificial intelligence is a branch of computer science. It is a field that combines computer science and robust datasets to enable problem-solving. It encompasses subfields, such as machine learning.

In the simplest explanation, AI is like teaching computers to think and act like humans. Just as we use our brains to solve problems, learn new things, and make decisions, AI helps computers do similar things. It is all about making machines smart enough to understand us, talk to us, and help us with tasks that normally need human smarts.

Basic AI Process

AI is a transformative field that harnesses systems and machines designed to replicate human intelligence, enabling them to execute various tasks. What sets AI apart is its capacity to learn and evolve through iterative processing and algorithmic training, much like how humans learn from experience. These systems continuously collect data, analyze patterns, and adjust their operations based on new inputs. In essence, AI systems have the unique ability to self-improve, becoming more proficient with each cycle of data processing.

The core mechanics of AI involve the amalgamation of extensive datasets with intelligent, iterative processing algorithms. These algorithms allow AI systems to discern patterns and features within the data they scrutinize. The advantage of AI is its ability to operate ceaselessly, around the clock, executing millions of tasks with remarkable speed. This constant activity allows AI to gather vast amounts of data, enabling it to learn, adapt, and enhance its capabilities over time.

As AI systems continue to evolve and accumulate knowledge, they contribute to the ongoing advancement of various fields and industries. Whether it is in healthcare, finance, or technology, the ability of AI to process data and learn from it has the potential to revolutionize how we approach complex problems and tasks, ultimately driving innovation and efficiency on a global scale.

AI Tasks

AI can perform a multitude of tasks, including automating repetitive processes, making predictions based on data analysis, recognizing patterns and anomalies, understanding and generating human language, interpreting visual information from images and videos, optimizing decision-making, assisting in medical diagnoses, recommending products and content, and simulating complex scenarios.

AI can also play strategic games, create art and music, translate languages, personalize user experiences, and enable autonomous systems in industries like transportation and robotics. AI's versatility and ability to learn and adapt from data make it a valuable tool for enhancing efficiency, accuracy, and innovation across a wide range of applications and industries.

Background

Birth of AI

Since early AI research was conducted in the 1950s, which explored topics like problem-solving, it has continued to evolve. In 1950, Alan Turing presented a paper that discussed how to build intelligent machines and test this intelligence. At the Dartmouth Workshop in 1956, where the birth of AI as a field is considered to have taken place, researchers gathered to explore the possibilities of creating machines that could simulate human intelligence. The first AI program was presented at the Dartmouth Summer Research Project on Artificial Intelligence (DSRPAI), which laid the foundation for AI research in the next few decades.

Early History

In the 1950s, early AI researchers focused on logic and symbolic reasoning to simulate human thought processes. Programs like "Logic Theorist" and "General Problem Solver" attempted to solve problems using rules and logic.

The 1960s–1970s period marked early successes and challenges. During this period, AI pioneers developed programs that could play chess, solve algebra problems, and understand natural language to a limited extent. The "Eliza" program simulated conversation, laying the foundation for chatbots. However, progress was slower than initially expected, and some early optimism gave way to an AI winter—a quiet period for AI research and development. In the 1960s, the US Department of Defense started to work in this field and began to train computers to mimic basic human reasoning.

In the 1970s, the Defense Advanced Research Projects Agency (DARPA) completed the street mapping project. The concept of expert systems was developed in the 1970s by Edward Feigenbaum, a computer scientist. An expert system is a computer program that uses AI technologies to simulate a human's judgment, behavior, and decision-making ability.

Progress: 1980s–1990s

The 1980s–1990s period saw the emergence of knowledge-based systems. Research shifted to knowledge-based systems, where AI systems used predefined rules and knowledge bases to make decisions. Expert systems like "MYCIN" diagnosed diseases, and "DENDRAL" identified chemical structures.

In the 1980s, AI development was boosted by expanding the algorithmic toolkit and more dedicated funds. Deep learning techniques were introduced, which enabled computers to learn through experience. In 1997, reigning World Chess Champion and Grandmaster Gary Kasparov was defeated by IBM's Deep Blue, a chess-playing computer program showcasing specialized AI's power in complex games.

Progress: Since the Turn of the Century

In the first decade of this century, machine learning gained prominence with improved algorithms and access to vast volumes of data. AI was integrated into online services like search engines, recommendation systems, and virtual assistants. In 2003, DARPA produced intelligent personal assistants, much before Siri or Alexa were developed and became household names.

In the next decade, deep learning was enabled by neural networks with many layers, which led to remarkable breakthroughs in image and speech recognition, natural language processing (NLP), robotic process automation, smart homes, and more. AI-powered products became mainstream, including virtual assistants like Siri and Alexa.

At this time, AI continues to advance in areas like self-driving cars, healthcare diagnostics, and creative tasks. In 2020, Baidu released the LinearFold AI algorithm to scientific and medical teams that were developing a vaccine during the early stages of the COVID pandemic. The algorithm could predict the RNA sequence of the virus in only 27 seconds, which is 120 times faster than other methods. Currently, research focuses on responsible AI, human-AI collaboration, and the challenges of AGI. During this period, ethical concerns around bias, transparency, and job displacement have become more prominent.

This timeline highlights key moments in AI history. However, it is important to note that AI's progress has been shaped by multiple phases of excitement, disillusionment, and resurgence, reflecting the challenges and opportunities of this dynamic field.

Drivers

A combination of technological advancements, societal needs, economic factors, incentives, and research breakthroughs has driven the development and popularity of AI.

The key drivers contributing to AI's growth interact and reinforce each other, leading to a dynamic cycle of innovation and progress.

Data, Performance, and Infrastructure Drivers

Data Explosion and Availability

Data plays a pivotal role in the advancement of AI, and the current landscape is characterized by the explosion of data and increased data availability. The proliferation of digital technologies has resulted in an unprecedented volume of data generated from diverse sources, including social media, sensors, and online activities. AI relies on massive data volumes to process its algorithms, learn, and make informed decisions.

This abundance of data has fueled the development of AI, supported by the emergence of various data labeling tools and the accessibility of structured and unstructured data storage and processing. The availability of vast amounts of structured and unstructured data, in turn, has empowered the creation of powerful machine learning models, underscoring the symbiotic relationship between AI and data.

Computational Power

Advances in hardware, particularly graphics processing units (GPUs) and specialized hardware like tensor processing units (TPUs), have significantly accelerated AI model training and inference processes and made complex computations feasible. It has also made it possible to train complex models faster and at larger scales, enabling the development of more sophisticated AI systems.

Affordable Computing Power

Affordable computing power has played a pivotal role in advancing artificial intelligence. With the ability to process massive volumes of data at a reasonable cost, AI development and usage have experienced a significant boost. This accessibility to high-performance computing resources has enabled researchers and businesses to harness AI's full potential, leading to innovations in various fields.

Algorithms and Models

Breakthroughs in developing sophisticated machine learning algorithms, particularly deep learning, have revolutionized AI. The development of advanced algorithms and techniques has enabled faster and more efficient processing that AI needs. Deep learning models can automatically learn complex, intricate patterns from data, leading to breakthroughs in tasks like language translation as well as image and speech recognition.

Cloud Computing

Cloud platforms have revolutionized the landscape of AI development by providing scalable and cost-effective computing resources. This transformation empowers both individuals and organizations to embark on AI projects without the need for substantial upfront investments in hardware infrastructure. With the flexibility and accessibility of cloud-based solutions, AI applications can be easily developed, deployed, and run, democratizing the field and fostering innovation across various sectors.

Open-Source Frameworks

The availability of open-source AI frameworks like TensorFlow and PyTorch has democratized AI development. These frameworks provide tools and libraries that researchers and developers can use to build AI models.

NLP Breakthrough

Advances in NLP have led to improved language understanding and generation by AI systems. It has enabled more human-like interactions between computers and humans, leading to the development of applications like chatbots, virtual assistants, and language translation services.

Business Drivers

Investment and Funding

The growing interest in AI has attracted increased investment from both public and private sectors, fueling research and development. This financial support enables the exploration of new ideas and technologies, advancing avenues for AI across research, development, and commercialization.

Industry Applications

AI has demonstrated its potential to improve efficiency and outcomes across healthcare, finance, manufacturing, and transportation industries. Hence, organizations are motivated to adopt AI to stay competitive and provide better services.

Competitive Advantage

AI provides previously unavailable insights and enables faster decision-making, providing enterprises with the competitive edge they seek. AI features and capabilities can lead to lower costs, reduced risks, faster time to market, and many other benefits.

Automation and Efficiency

AI-driven automation is a compelling solution for businesses, as it has the potential to streamline operations, cut down on expenses, and significantly boost productivity across diverse industry domains. Its ability to handle repetitive tasks efficiently while continuously learning and adapting makes it an indispensable tool for organizations seeking operational optimization.

Robotics and Autonomous Systems

AI has found applications in robotics and automation, leading to the development of self-driving cars, drones, and industrial robots that can perform tasks with minimal human intervention. The integration of AI with robotics has transformed industries and created new possibilities.

Healthcare and Medicine

AI has shown potential to enhance medical diagnosis, drug discovery, image analysis, and personalized treatment plans, improving patient outcomes and reducing healthcare costs. These applications have driven interest and investment in AI for healthcare.

Energy Efficiency and Sustainability

AI holds immense potential for bolstering sustainability initiatives through its capacity to optimize energy usage resource allocation and tackle environmental issues head-on. By leveraging AI technologies, organizations can achieve more efficient operations while simultaneously contributing to a greener, more sustainable future.

Global Collaboration and Research

The international AI research community collaborates to share knowledge, publish findings, build on each other's work, and drive innovation. This collaboration has accelerated progress and innovation in AI. Global initiatives focus on addressing challenges like ethics, fairness, and safety in AI.

Business and Industry Demand

Industries are adopting AI to improve efficiency, automate tasks, and gain competitive advantages. From finance to healthcare, AI is transforming the way businesses operate.

Consumer Demand

The widespread availability of AI-driven consumer products and services, such as smart assistants and recommendation systems, has ignited a surge in interest and subsequent adoption of AI technology. These consumer-oriented AI applications have improved convenience and demystified and popularized the use of AI in everyday life, making it more accessible to a broader audience.

National Strategies

Numerous countries and governmental bodies have acknowledged the strategic significance of AI. In response, they have not only devised comprehensive national policies and strategies but have also initiated efforts to foster research, development, and the integration of AI technologies. Moreover, these entities have launched specific initiatives geared toward supporting research and development in AI and initiatives aimed at enhancing education in the field.

Ethical and Social Concerns

The ascent of AI has ignited conversations around ethical concerns, drawing attention to issues like bias, transparency, and accountability within AI systems. This increased emphasis on responsible AI has been instrumental in driving efforts to create fair and unbiased AI technologies, ensuring that they benefit society while minimizing potential harm.

Educational Resources

With the proliferation of online courses, tutorials, and educational platforms, individuals now have convenient access to resources that enable them to learn about AI and acquire the necessary skills to engage with AI technologies. These accessible learning opportunities have democratized AI education, making it possible for a wider range of people to enter the field and contribute to its advancement.

2

Benefits and Disadvantages

AI Applications Across Industries

AI has spearheaded a transformative wave across diverse industry sectors, reshaping the landscape of applications. In manufacturing, the integration of AI-powered robotics and automation systems has ushered in a new era of efficiency and precision in production processes. AI's predictive maintenance capabilities, driven by sophisticated algorithms, play a pivotal role in minimizing downtime by anticipating equipment failures before they occur, thereby enhancing overall operational reliability.

Financial institutions leverage AI for myriad purposes, including fraud detection, risk assessment, and algorithmic trading. The implementation of AI-driven solutions contributes to more robust security measures and informed decision-making in the financial domain. Moreover, the realm of customer service has been revolutionized by deploying chatbots and virtual assistants, ensuring instant and personalized responses to customer inquiries.

Transportation and logistics stand out as another sector where AI applications bring substantial advantages. AI-driven algorithms optimize routes, reducing fuel consumption and improving overall operational efficiency. These instances vividly illustrate how AI serves as a catalyst, augmenting productivity, accuracy, and decision-making capabilities across various industries, setting the stage for unprecedented levels of innovation and efficiency.

The next section delves into specific examples, providing a closer look at how AI is harnessed across diverse domains, delivering tangible benefits in myriad ways.

A. Khan, *Artificial Intelligence: A Guide for Everyone*,
https://doi.org/10.1007/978-3-031-56713-1_2

Benefits Across Applications

Manufacturing and Operations

AI plays a crucial role in both manufacturing and quality control processes. In manufacturing and operations, AI-powered robots and automation systems contribute to increased efficiency and improved quality control. These technologies optimize production processes, enhance maintenance scheduling, minimize unplanned downtime, and elevate overall equipment efficiency. Simultaneously, in quality control and inspection, AI systems conduct real-time inspections, swiftly identifying defects and inconsistencies in materials, components, and finished products. This proactive approach improves manufacturing processes and boosts customer satisfaction by reducing defects and minimizing waste.

Automation and Robotics

AI significantly contributes to automation across various domains, including industrial tasks and transportation. In manufacturing, AI-driven robots execute tasks with precision, while in transportation, AI plays a pivotal role in the development of autonomous vehicles. Additionally, AI powers robotic process automation (RPA) to automate routine business tasks, streamline operations, and improve efficiency. This makes AI indispensable in modern workplaces, automating tasks, reducing human intervention, and boosting overall productivity.

Accuracy and Consistency

AI systems exhibit remarkable capabilities in carrying out tasks with exceptional precision and consistency, effectively mitigating the risk of errors that can result from human fatigue or oversight. This reliability has positioned AI as a valuable tool in various industries, where accuracy and consistency are paramount, from manufacturing to healthcare and beyond.

24/7 Availability

AI-powered systems are designed to operate tirelessly without the necessity for breaks or rest, ensuring uninterrupted service and support, which can be particularly advantageous in scenarios requiring 24/7 availability. This continuous operation enhances efficiency and responsiveness, contributing to improved productivity.

Speedier Processes

AI is pivotal in expediting the optimization and improvement of product, manufacturing, and business processes, significantly accelerating the pace at which innovations and efficiency gains can be achieved.

Workload Scalability

AI solutions offer the advantage of scalability, allowing them to seamlessly adapt to handle large workloads and shifting requirements, making them a flexible choice for businesses experiencing growth or evolving demands. This scalability ensures that organizations can continue to leverage AI effectively as their operations expand and evolve over time.

Remote Monitoring

AI-enabled sensors and devices can monitor and analyze data remotely, offering significant improvements in managing various critical infrastructures, machinery, and complex systems. This technology enables real-time insights, predictive maintenance, and increased efficiency in a wide range of industries.

Safety in Hazardous Environments

Robots and drones driven by AI capabilities excel in executing operations within hazardous settings, effectively minimizing human exposure to potentially life-threatening dangers. This technology contributes to the safer and more efficient execution of tasks in environments where human intervention might be perilous.

Supply Chain Management

AI revolutionizes manufacturing supply chains by optimizing production schedules, managing inventory effectively and accurately forecasting demand.

Its predictive analytics enhance operational efficiency, ensuring optimal inventory levels and minimizing disruptions.

Healthcare

AI is indispensable in healthcare, playing a vital role in medical diagnosis by analyzing diverse data sources for precise disease identification. Additionally, it accelerates drug discovery through analytical capabilities, expediting the identification of potential drug candidates and advancing pharmaceutical research. This transformative impact of AI in healthcare is further explored in the following subsections.

Diagnostics and Imaging

AI plays a pivotal role in healthcare, specifically in diagnostics, medical imaging analysis, and disease prediction. Medical professionals benefit from AI's assistance in accurate and efficient disease diagnosis based on medical images. Notably, tools like IDx-DR detect conditions such as diabetic retinopathy from retinal images, showcasing AI's potential to improve patient outcomes.

Disease Prediction

AI is pivotal in disease prediction, leveraging vast datasets to identify patterns and correlations within medical records, genetic information, and lifestyle factors. In healthcare, AI-driven predictive models offer valuable insights into disease likelihood, enabling early intervention and personalized preventive strategies. This transformative application fosters proactive healthcare, shifting from reactive treatment to proactive health management. Predictive analytics, including tools like Deep Patient, enhances healthcare by identifying risks and recommending personalized treatments based on a patient's medical history.

Drug Discovery

AI has revolutionized drug discovery by expediting the identification of potential compounds and accelerating the research and development process. AI algorithms analyze massive datasets, including chemical structures, biological interactions, and existing drug databases, to predict potential drug candidates. This data-driven approach significantly enhances the efficiency of screening processes, reducing the time and resources required for drug discovery. By uncovering novel insights and patterns within complex datasets, AI facilitates the identification of promising compounds for further investigation, ultimately contributing to the development of innovative and life-saving medications.

Business Applications

AI has value for almost every business function in most industries. It significantly impacts various aspects of enterprise operations, transforming how businesses operate, make decisions, and engage with customers. The following are some key areas where AI is used in the enterprise.

Data Analysis

AI-driven data analysis automates the processing of large volumes, revealing actionable insights crucial for strategic decisions. Using machine-learning models leveraging historical data, it predicts trends, customer behavior, and market dynamics. Rapid and accurate processing of massive datasets, both human-generated and machine-generated, enables unparalleled understanding, surpassing human capacity. AI maximizes data utilization, extracting insights that might elude human observation and facilitating data-driven predictions for informed business decisions.

Decision-Making

Decision-making in today's data-rich environment poses challenges due to the sheer volume of data from diverse sources, surpassing human capacity for absorption and interpretation. AI systems play a pivotal role in this context, facilitating data-driven insights and significantly enhancing decision-making across diverse business domains. By swiftly processing vast datasets, AI systems not only provide real-time decision-making capabilities but also ensure accuracy and efficiency. This is particularly crucial in applications like fraud detection, where timely responses and instantaneous data analysis are imperative to identify and mitigate potential risks. The integration of AI into decision-making processes contributes to a more streamlined, adaptive, and effective approach.

Human Resources

AI is transforming human resources (HR) by automating various tasks, from candidate sourcing to resume screening, saving valuable time for HR professionals. Machine-learning algorithms enable the identification of candidates that match specific criteria efficiently. Innovative interview tools powered by AI assess candidate responses, offering insights to support effective hiring decisions and enhancing the overall recruitment process. Beyond recruitment, AI extends its impact to employee engagement and sentiment analysis tools, providing HR professionals with valuable data to monitor and improve workplace satisfaction. This integration of AI in HR processes contributes to increased efficiency, objectivity, and precision, ultimately fostering a positive work environment and supporting the well-being of the workforce.

Finance

AI is transforming finance, particularly in algorithmic trading and fraud detection. In trading, advanced algorithms enable rapid financial decisions and swift trade execution. AI analyzes transaction data for fraud detection to prevent illicit activities, ensuring financial market integrity. Additionally, AI plays a crucial role in risk management by assessing market dynamics, improving customer service through chatbots, and aiding portfolio management by creating diversified investment portfolios. Analyzing financial data, AI detects anomalies, fraudulent activities, and market trends, enhancing security. Algorithmic trading leverages AI for quick, data-driven investment decisions. Chatbots assist with customer inquiries, and robo-advisors provide automated investment advice, promising more efficient, secure, and responsive financial services.

Customer Applications

Technology converges with personalized experiences in AI customer applications, transforming how businesses interact with their clientele. This section explores the myriad ways AI is revolutionizing customer engagement and satisfaction, from advanced chatbots to tailored recommendation engines.

Customer Service and Support

AI-driven chatbots transform customer support, delivering instant assistance and minimizing wait times for enhanced satisfaction. For instance, companies like H&R Block utilize IBM's Watson Assistant for specialized tax-related support. These chatbots excel in understanding queries, providing relevant information, and guiding users through complex processes. Beyond support, AI fosters efficient relationships with customers and partners, offering prompt responses to enhance user satisfaction. The 24/7 customer service includes FAQ responses and routine inquiries on websites and apps. Virtual assistants handle routine tasks, freeing human agents for more complex issues and optimizing overall support systems.

Language-Related Capabilities

AI excels in language-related capabilities, including accurate language translation, sentiment analysis to discern emotional tones, and proficient speech recognition for seamless voice-to-text transcription. AI-driven chatbots and virtual assistants engage in real-time conversations, addressing queries and enhancing user interactions. These language translation tools play a crucial role in fostering effective communication across diverse linguistic backgrounds, breaking down barriers, promoting global connectivity, and facilitating cross-cultural exchanges. These AI-driven tools ensure accurate and efficient language translation, contributing to cross-cultural understanding and collaboration in our interconnected world and offering real-time translation capabilities.

Recommendation Systems

AI's transformative impact extends to the digital realm, particularly e-commerce and content recommendation. AI suggests tailored products, articles, videos, and various content types, leveraging user preferences and browsing history, delivering a personalized digital experience across platforms. In e-commerce, AI, exemplified by Amazon's recommendation engine, enhances user experience and drives sales. This influence extends to marketing campaigns, product recommendations, and content delivery, optimizing communication and content through AI in marketing automation platforms. AI's analytical capabilities, grasping user behavior, guarantee personalized content, enhancing the overall user experience.

Enhanced User Interfaces

Through natural language interactions and gesture recognition, AI enables the creation of user interfaces that are more intuitive and user-friendly, elevating the overall experience for individuals using a wide range of applications and devices. These technologies pave the way for seamless and user-centric interactions, making technology more accessible and user-friendly.

Virtual Assistants

Virtual assistants powered by AI, such as Siri, Alexa, and Google Assistant, offer swift and convenient access to information, streamline various tasks, and play a supportive role in daily activities, making them valuable tools in our digitally connected lives. Their ability to understand and respond to voice commands has transformed the way we interact with technology, enhancing our productivity and convenience.

Education

AI plays a pivotal role in revolutionizing education through personalized learning experiences tailored to individual student needs and learning preferences. Serving as a versatile tutor, AI offers invaluable assistance and detailed explanations across various subjects, promoting enhanced comprehension and mastery of educational materials. This transformative approach ensures that education is adaptive, accessible, and effective, fostering lifelong learning and academic success. AI-driven e-learning platforms further contribute to this revolution by adapting content and pacing to individual needs, providing timely feedback, and fostering a dynamic learning environment. Integrated chatbots serve as virtual assistants, readily available to address course-related questions and support students throughout their educational journey.

Gaming

AI profoundly impacts the gaming industry, enhancing player experiences through several key functions. Non-player characters (NPCs) benefit from AI by exhibiting intelligent behavior, making gameplay more immersive and challenging as these characters interact with players. Additionally, AI excels in procedural content generation, creating dynamic game levels, characters, and environments, ensuring that no two gaming experiences are identical. These AI-driven capabilities contribute to the diversity and excitement of modern video games, providing gamers with engaging and ever-evolving content.

Commerce

AI revolutionizes operations from e-commerce enhancements to sales analytics, marketing strategies, customer spending analysis, and price optimization. It has become a driving force behind efficiency and strategic decision-making, reshaping the landscape of modern business, as described in the following subsections.

E-commerce and Retail

In e-commerce and retail, sophisticated recommendation engines utilize AI algorithms to analyze customers' browsing and purchase history, providing personalized product suggestions. Moreover, AI contributes to more efficient inventory management, leveraging advanced analytics for precise demand forecasting. Additionally, pricing optimization strategies are enhanced through AI, ensuring businesses can adapt dynamically to market trends and customer behaviors.

Sales and Lead Generation

AI tools analyze customer interactions, sales data, and historical trends to identify potential leads and prioritize sales efforts. AI-driven sales assistants provide sales teams with real-time insights, suggestions, and reminders to improve their interactions with customers.

Marketing

AI-driven marketing automation helps businesses personalize marketing campaigns, segment customers, and optimize ad targeting for better ROI. Predictive analytics is used to identify potential leads and prioritize sales opportunities.

Customer Spend

AI leverages transactional and demographic data to forecast the anticipated spending of specific customers throughout their interactions with a business, commonly referred to as customer lifetime value (CLV). This predictive modeling assists companies in tailoring their strategies to enhance customer relationships and optimize revenue generation.

Price Optimization

AI-driven price optimization involves using advanced algorithms and machine learning to analyze market dynamics, customer behavior, and various external factors to determine the most optimal pricing strategy for products or services. This approach helps businesses maximize revenue, improve competitiveness, and adapt to market changes with data-driven pricing decisions.

Technology and Innovation

AI offers a wide array of advantages across various domains, including streamlining map directions, enabling seamless mobile banking experiences, optimizing smart homes for greater convenience, and enhancing investment analysis for informed financial decisions. Its versatility and integration into daily life underscore the transformative impact of AI technologies, as described in the following sections.

Autonomous Vehicles

AI is at the forefront of autonomous vehicles, driving innovations in self-driving cars and drones. Waymo, a subsidiary of Alphabet Inc., exemplifies this with its AI-powered self-driving car technology, which navigates roads, interprets traffic signals, and adapts to dynamic environments, thereby reducing the risk of human error in driving. The development of self-driving cars and drones underscores the pivotal role of AI in mitigating accidents caused by human errors and enhancing the overall efficiency of transportation systems. This advancement holds significant potential to revolutionize mobility, making it safer.

Smart Cities

AI is harnessed to revolutionize urban infrastructure by optimizing traffic management, conserving energy resources, and improving public safety through enhanced surveillance and predictive policing techniques. These applications contribute to more efficient and sustainable city living, benefiting both residents and the environment.

Environmental Management

AI plays a multifaceted role in environmental sustainability, optimizing energy consumption in buildings and industrial processes to reduce costs and mitigate environmental impact. Additionally, AI contributes to environmental monitoring by analyzing sensor data to manage air and water quality, thereby preserving ecological balance. These initiatives underscore AI's instrumental role in fostering sustainable practices, reducing resource wastage, and safeguarding the environment. The revolution in energy management and the application of predictive maintenance systems further highlight AI's pivotal role in advancing sustainability and enhancing critical infrastructure efficiency.

Agriculture

AI analytics and remote-sensing technologies have revolutionized agriculture by enhancing crop monitoring, accurately predicting yields, and optimizing resource allocation for more sustainable and efficient farming practices. These precision agriculture techniques not only increase productivity but also minimize waste and environmental impact, making them valuable tools for modern farming.

Scientific Research

AI is a cornerstone in scientific research, aiding data analysis across diverse domains like astronomy, genomics, and particle physics. Its computational power and pattern recognition capabilities empower researchers to uncover insights, identify trends, and make sense of intricate data, accelerating scientific discovery in specialized fields. Additionally, AI plays a pivotal role in exploring remote and hazardous terrains where human intervention is challenging. These AI-powered systems navigate and investigate such environments, providing invaluable insights that bolster scientific research and contribute to judicious resource management.

Miscellaneous

Search and Information Retrieval

AI powers the core functions of search engines, utilizing sophisticated algorithms to deliver highly relevant search results tailored to user queries. Beyond search, AI excels in information extraction, efficiently gathering pertinent data from documents, websites, and diverse sources. These capabilities empower users to access precise information swiftly and enhance data-driven decision-making, making AI an indispensable tool for information retrieval and knowledge management in our digital age.

Creativity

AI-powered applications revolutionize entertainment by reshaping the creative process, spanning from art and music generation to various imaginative content production. These tools not only redefine creativity and productivity but also open new possibilities for artistic expression and collaboration across diverse domains. Evident in art, music, and writing, AI's creative potential enriches the artistic experience, offering innovative solutions for content creation and information generation. This transformative impact underscores AI's role in pushing the boundaries of human creativity and enhancing the overall creative landscape.

Content Generation and Moderation

AI's prowess in writing and content generation transforms creativity and productivity, producing diverse written content efficiently. In publishing, The Associated Press, utilizing AI software, achieved remarkable efficiency gains, generating 12 times more stories by automating short earnings news stories. This streamlined process allows journalists to focus on crafting in-depth pieces. AI's evolution includes automating content-moderation tasks for online safety and revolutionizing digital information production and consumption.

Economic Growth

AI technologies are catalysts for innovation, fostering the emergence of new business opportunities and industries that, in turn, drive economic growth and provide fresh avenues for entrepreneurial ventures. Their disruptive potential spans various sectors, stimulating progress and reshaping the economic landscape.

Cost Savings

The integration of AI into business operations results in significant cost savings as it streamlines processes, eliminates inefficiencies, reduces waste, and ultimately trims down operational expenses. AI's enhanced automation and efficiency not only improve the bottom line but also enhance overall competitiveness in today's fast-paced business environment.

Risk Management

AI excels in assessing and predicting real-time risks by analyzing a diverse range of data sources, empowering businesses to make well-informed decisions that can effectively mitigate potential losses. This dynamic risk analysis capability enhances risk management strategies and supports proactive decision-making in various industries.

Innovative Research

By swiftly analyzing intricate datasets and running simulations, AI accelerates scientific research, shortening the time required to make critical discoveries and advancements. This powerful technology aids scientists in solving complex problems, unlocking new insights, and pushing the boundaries of human knowledge.

Legal and Compliance

AI plays a pivotal role in various aspects of the legal domain, including aiding in legal research, contract analysis, and due diligence procedures, significantly improving efficiency and accuracy. Additionally, compliance-monitoring tools equipped with AI capabilities are instrumental in proactively identifying potential regulatory breaches, helping organizations maintain adherence to legal and industry standards.

Disadvantages

In the ever-evolving landscape of artificial intelligence, a comprehensive examination of its limitations becomes essential. While AI has undeniably showcased unprecedented capabilities across various domains, ranging from task automation to decision-making support, its deployment presents a myriad of challenges and ethical considerations. This section serves as a nuanced exploration of the disadvantages associated with AI technologies, aiming to illuminate issues and the ethical quandaries arising from AI's expanding role in our lives.

Reliance on Data

AI systems rely extensively on substantial volumes of high-quality data to undergo effective training and make informed decisions. The absence of adequate data or the presence of biased datasets can act as substantial obstacles, compromising the performance and reliability of AI systems. Thus, data quality and diversity remain pivotal factors in enhancing the capabilities and fairness of AI technologies across various applications and industries.

Complexity and Dependence

Overreliance on AI systems can lead to complex systems that are difficult to maintain, understand, or troubleshoot. Humans can become overly reliant on AI. Relying on AI for critical tasks can create vulnerabilities if the system malfunctions, leading to potential disruptions or failures.

False Sense of Security

The remarkable capabilities of AI can occasionally foster a misleading sense of security, resulting in unwarranted trust in systems that are not infallible. This overreliance on AI can have consequences, especially when dealing with critical decisions or situations where human oversight and judgment are indispensable. Therefore, it is imperative to maintain a balanced approach, acknowledging AI's strengths while remaining vigilant and discerning in its application.

Unpredictability

Deep-learning neural networks, among other complex AI models, are known for their capacity to yield results that can be challenging to explain or understand, leading to unpredictability in their decision-making processes. This lack of transparency poses a significant challenge in ensuring the accountability and interpretability of AI systems, warranting the development of methods to make these models more understandable and trustworthy.

Misuse and Manipulation

AI, while a powerful tool, also carries the potential for malicious use, as it can be harnessed to generate deepfake content, orchestrate cyberattacks, and propagate false information, posing significant risks to society and cybersecurity. Safeguarding against these threats requires a comprehensive approach, including responsible AI development, regulatory measures, and ongoing monitoring of AI systems to detect and mitigate misuse.

Environmental Impact

The computational requirements of training and operating AI models can result in a notable carbon footprint, raising concerns about their environmental impact. Addressing this challenge necessitates the development of more energy-efficient AI algorithms and the adoption of sustainable practices in AI infrastructure to minimize the environmental consequences of AI technology.

Legal and Regulatory Challenges

AI technology introduces legal and regulatory complexities, particularly concerning liability issues, especially when AI-driven errors or accidents occur. Navigating these challenges necessitates the development of clear legal frameworks and policies to determine responsibility and accountability in cases involving AI systems.

Ethical Concerns

AI systems can inadvertently perpetuate biases present in the training data, leading to unfair and discriminatory outcomes. As AI advances, there are long-term ethical concerns related to the potential development of advanced AI systems, such as artificial super intelligence (ASI), which raise questions about control and safety.

Lack of Creativity and Intuition

While exceptionally capable in specific domains, AI falls short of human-level creativity, intuition, and emotional comprehension, thereby restricting its effectiveness in managing intricate, unstructured tasks that require a deep understanding of human emotions and nuanced contexts. These limitations underscore the importance of human–AI collaboration, where AI can complement human strengths, but human judgment remains crucial in certain scenarios.

Availability of Practical Products

Currently, there are few practical AI products widely available for everyday use. However, there is a growing anticipation that this landscape will undergo a substantial transformation in the near future, with AI becoming more integrated into various aspects of our lives and industries. The rapid development and deployment of AI solutions are poised to make AI a common and indispensable part of our daily routines.

Cost and Infrastructure

The development and implementation of AI systems often demand significant investments, including high-performance computing infrastructure and recruiting skilled personnel with expertise in machine learning and data science. These investments are crucial for creating and maintaining effective AI solutions, but they can provide substantial returns in terms of efficiency and innovation.

Addressing Shortcomings

It is important to note that while AI offers many benefits, it also comes with certain disadvantages and challenges that need to be carefully considered. Responsible implementation of AI is essential to ensure that its benefits are realized without negative impacts and that its benefits are maximized while mitigating potential risks. These disadvantages can be addressed through responsible AI development, ongoing research, regulatory frameworks, and ethical considerations, ensuring that AI's potential benefits are maximized while minimizing its potential negative impacts.

3

AI–Human Relationship

Navigating the AI–Human Landscape

Evolving AI–Human Dynamics

The multifaceted nature of the relationship between AI and humanity has provided adaptability and versatility, reshaping industries and revolutionizing healthcare. This intricate tapestry is woven from myriad threads, each representing a unique facet of interaction. Beyond a simplistic man-versus-machine narrative, complex layers define this engagement, transcending binary perspectives.

AI is a tool and collaborator, influencing how we work and think. It operates in technological, psychological, societal, and ethical dimensions, fostering adaptability and reshaping industries. This collaborative synergy illustrates the symbiotic evolution between AI and human roles. Recognizing it as an evolving phenomenon requires a nuanced understanding that moves beyond the binary discourse of AI as either a threat or a savior.

As our world becomes increasingly AI-driven, an intricate and multifaceted relationship exists between AI and humanity. AI is not a distant force but an integral part of our contemporary reality, influencing decision-making processes, shaping industries, and redefining social interactions. Therefore, using a holistic view will enable informed judgments about the role of AI in our lives and society.

A. Khan, *Artificial Intelligence: A Guide for Everyone*,
https://doi.org/10.1007/978-3-031-56713-1_3

Context and Significance of the AI–Human Relationship

The intertwining relationship between AI and humanity is a defining charac-teristic of our contemporary technological landscape. As we enter an AI-driven era, exploring the nuanced dynamics that shape this interaction becomes imperative. Rapid advancements in AI technologies, from machine learning to natural language processing, have permeated various aspects of our lives, raising questions about human identity, labor, and societal structures. Understanding the context and significance of the AI–human relationship is crucial for thoroughly examining the symbiotic interaction between AI and humanity.

Interaction's Multifaceted Nature

The AI–human relationship is intricate, woven with diverse elements, each portraying a unique facet. Beyond a simplistic man-versus-machine narrative, intricate layers define this engagement. AI influences how we work and think. It surpasses binary perspectives, encompassing technological, psychological, societal, and ethical dimensions.

The complexity fosters adaptability, reshaping industries and revolutioniz-ing healthcare. Recognizing it as an evolving phenomenon requires a nuanced approach to understanding. The intricate relationship brings adaptability and versatility, illustrating a collaborative synergy in the symbiotic evolution between AI and human roles.

Reciprocal Influences

Interplay of AI and Human Dynamics

The dynamic interplay between AI and human dynamics characterizes our contemporary technological landscape. This intricate relationship transcends a simple dichotomy, recognizing AI not as an external force but as a dynamic collaborator shaping human experiences. As AI integrates into our daily lives, influencing work, communication, and perception, a symbiotic nature becomes evident.

Humans are influenced by technological advancements, creating a recipro-cal relationship across societal structures and individual experiences. Moving beyond the narrative of AI solely as a threat or savior, this interplay acknowledges multifaceted roles, emphasizing adaptability and ethical considerations in navigating the evolving terrain.

The dynamic interplay underscores the importance of fostering a holistic understanding, highlighting that the relationship is not static but continuously evolving. As AI and humans shape each other's trajectories, adaptability, understanding, and ethical considerations become essential in navigating this dynamic terrain. The evolving relationship encourages a responsible and beneficial integration of AI technologies, aiming to align with human values and contribute positively to collective progress.

Mutual Influence of AI and Human Dynamics

The interdependence between AI and human dynamics creates a reciprocal influence that significantly shapes the landscape of our technological interactions. As AI evolves, it adapts to human behaviors, learning from patterns and responses. This adaptive capability enables AI systems to provide more personalized and efficient solutions in customer service, recommendation algorithms, or other interactive domains. Simultaneously, human dynamics influence AI development through the ethical considerations, preferences, and societal values integrated into the design and deployment of AI technologies. The collaboration between AI and human dynamics leads to a continuous feedback loop, where advancements in technology prompt changes in human behavior, and evolving human needs, in turn, steer the direction of AI innovation.

The impact of AI on human dynamics extends beyond efficiency gains. It influences decision-making processes, redefines job roles, and shapes societal paradigms. As AI becomes integral to our daily lives, from intelligent assistants to predictive analytics, individuals and communities adapt to these technological shifts. The symbiotic relationship between AI and human dynamics emphasizes the need for thoughtful consideration of ethical implications, privacy concerns, and the broader societal consequences of technological advancements. In navigating this interplay, it becomes essential to strike a balance that aligns AI development with human values and societal well-being, fostering a harmonious integration that contributes positively to our collective progress.

Relationship with Humans

Automation and Enhancement in Human Roles

AI presents a dual prospect by offering the potential to automate and enhance tasks that humans have traditionally carried out. This dynamic interplay between AI and human roles introduces a complex and multifaceted dimension to the impact of AI on employment. While it is undeniable that AI can automate and optimize various tasks, the idea that it will completely replace humans in every facet of life remains unlikely.

Understanding the nuanced impact of AI on human roles requires considering several factors. Rather than solely focusing on the prospect of AI replacing human workers, it is crucial to recognize that AI primarily revolves around the process and capability for superpowered thinking and advanced data analysis. Beyond the realm of automation and robots, the overarching goal of AI is to augment and enhance human capabilities. This augmentation is envisioned as a valuable asset, particularly in business operations.

AI in Daily Lives

Integrating AI into our daily lives and societal structures has become increasingly pervasive, fundamentally altering how we live, work, and interact. From personalized recommendations on streaming platforms to voice-activated virtual assistants, AI has seamlessly woven into the fabric of our routines. It plays a pivotal role in healthcare through diagnostic tools and treatment advancements, transforms education with personalized learning experiences, and revolutionizes industries through automation and predictive analytics.

Societal Structures

Beyond individual experiences, AI influences societal structures by reshaping decision-making processes, labor markets, and even the dynamics of social interactions. The impact extends to ethical considerations, privacy concerns, and the redefinition of traditional norms. Acknowledging AI's role in daily life and societal structures requires a holistic understanding that embraces both its opportunities and challenges, emphasizing the need for ethical frameworks and thoughtful considerations as we navigate the evolving landscape of this transformative technology.

Critical Areas of AI Impact

Task Automation

Task automation is one of the cornerstones of AI's transformative impact, particularly in industries where routine and repetitive tasks are prevalent. From manufacturing to data entry and customer support, AI showcases its proficiency in streamlining operations and enhancing productivity. AI-driven robotic systems have become instrumental in introducing precision to complex assembly processes in manufacturing. This reduces errors and significantly elevates production efficiency, leading to higher output quality. It is essential to note that while the automation of predictable and monotonous tasks may impact specific job roles, the overarching narrative isn't one of total job displacement. Instead, automation often catalyzes role transformation, fostering a collaborative dynamic where human workers complement AI systems, leveraging their strengths for more efficient and innovative outcomes.

Beyond the immediate impact on job roles, integrating AI in task automation brings a paradigm shift in work. It allows human workers to redirect their focus toward more creative, strategic, and complex aspects of their roles while AI handles routine functions. This collaboration between humans and AI redefines job responsibilities. It opens avenues for upskilling and continuous learning, ensuring that the workforce remains adaptive and resilient despite evolving technological landscapes.

Augmenting Human Capabilities

The augmentation of human capabilities through collaboration with AI unveils powerful synergies that redefine the landscape of various industries. AI plays a crucial role by providing valuable tools and insights that enhance decision-making, particularly in healthcare. It supports medical professionals in diagnosing diseases through advanced data analysis and image recognition, contributing to more accurate and timely medical decisions. This collaborative relationship is characterized by a blend of cooperation, adaptation, and ongoing learning, navigating the dynamic intersection of work and technology.

Within this partnership, AI excels at repetitive tasks, data analysis at scale, and pattern recognition within vast datasets. Human professionals contribute critical thinking, ethical judgment, and nuanced context understanding, particularly in tasks involving empathetic patient interaction in healthcare. This dynamic interplay enhances decision-making, problem-solving, and productivity across industries, emphasizing the value of combining AI's computational prowess with human cognitive and ethical expertise. AI is not viewed as a replacement but as a potent tool, serving as a valuable ally to augment human capabilities. This collaborative synergy represents a fundamental shift in the workforce, underlining the transformative potential of this partnership in shaping the future of work and productivity.

Moreover, the evolving partnership between humans and AI opens up new possibilities for innovation and creativity. As AI takes over routine tasks, human professionals are freed up to focus on more complex and imaginative aspects of their work. This shift encourages a reimagining of job roles and responsibilities, fostering an environment where human ingenuity and creativity can flourish. The coexistence of AI and human skills creates a harmonious balance that improves efficiency and lays the foundation for groundbreaking advancements in various fields. As industries adapt to this collaborative future, the synergy between human intelligence and AI capabilities promises to redefine the nature of work and productivity.

Jobs

The continuous evolution of AI technologies transforms existing job roles and creates novel opportunities in specialized fields. Professions in AI development, data analysis, machine learning engineering, and AI ethics emerge as key domains where human expertise is indispensable. These fields underscore human professionals' intricate and nuanced role in creating, maintaining, and enhancing AI systems. While AI showcases proficiency in specific tasks, it currently grapples with endeavors that demand intricate problem-solving, creative thinking, emotional intelligence, and nuanced commonsense reasoning.

The ongoing synergy between AI and human cognitive abilities highlights the resilience of certain job roles against automation. Professions that heavily rely on creativity and emotional intelligence, such as artists, writers, psychologists, and strategic decision-makers, continue to thrive due to the unique capacities humans bring to these roles. Similarly, complex and dynamic positions in fields like law, medicine, and research necessitate human expertise, given their tasks' nuanced and context-dependent nature. This dual perspective underscores the dynamic nature of contemporary job roles in an AI-driven era, where human involvement and specialization remain enduringly significant, shaping the evolving landscape of the workforce.

Impact on Industries

The influence of AI manifests as a nuanced landscape across diverse industries, showcasing varying degrees of impact. Sectors such as manufacturing and customer service stand at the forefront of significant automation, where AI-driven tools and robotics streamline processes, optimize efficiency, and transform traditional workflows. The integration of AI in these domains reflects a commitment to technological advancements that enhance productivity, reduce errors, and propel industries toward a more automated future.

Conversely, industries like healthcare and creative arts place a premium on the irreplaceable domain of human expertise. In healthcare, the nuanced interplay of empathy, intricate decision-making, and personalized care remains central to core functions, establishing a realm where AI acts as a supportive tool rather than a replacement for human professionals. Similarly, the distinctive touch of artistic creativity and emotional intelligence in creative arts distinguishes human contribution from AI capabilities. This divergence underscores the tailored application of AI, emphasizing its capacity to augment human capabilities and the importance of recognizing the distinctive needs and dynamics within each industry as it navigates the evolving AI landscape.

Cognitive and Creative Tasks

AI exhibits remarkable proficiency in specialized tasks, excelling in pattern recognition and data analysis. However, it fails to replicate the broader spectrum of human traits, including general intelligence, creativity, intuition, and emotional understanding. This limitation positions job roles dependent on complex decision-making, critical thinking, creativity, and empathy as less susceptible to full automation. Instead, these professions are poised for transformation, envisioning AI as a complementary tool that empowers human professionals.

In this evolving landscape, AI serves as a catalyst for innovation, offering valuable support to human experts in harnessing data-driven insights, streamlining processes, and making informed decisions. The synergy between AI and human capabilities becomes evident as professionals leverage AI to augment their skills, creating a dynamic collaboration that capitalizes on the strengths of both. This paradigm underscores the enduring value of uniquely human attributes in a technology-driven world, emphasizing the coexistence and mutual reinforcement of AI and human intelligence in shaping the future of cognitive and creative tasks.

Ethical and Social Implications

The integration of AI into society presents a multitude of ethical and societal considerations. One major concern is the potential for bias within AI systems, perpetuating and exacerbating existing inequalities. Discussions on the possibility of job displacement due to automation have prompted considerations for reskilling and adapting the workforce for an AI-driven future. Additionally, the ethical ramifications of data privacy and broader societal implications on social dynamics, including human–AI interactions, continue to be subjects of critical scrutiny and debate as AI technologies evolve.

The integration of AI into society is deeply intertwined with ethical, moral, and social considerations that significantly impact its adoption. Human values often influence decisions regarding AI usage, encompassing concerns related to fairness, accountability, and transparency. Recognizing the need for human oversight in AI systems to ensure ethical and responsible use is paramount. This underscores the intricate interplay between technology and human values, where the moral dimension of AI serves as a guiding compass, steering its development and deployment toward a future where technology aligns with the betterment of humanity.

Moreover, the ethical implications extend to questions of autonomy as AI systems become increasingly sophisticated and capable of autonomous decision-making. Striking a balance between the autonomy of AI and the ethical principles guiding its actions is essential. Discussions surrounding AI ethics delve into the moral responsibility of developers, policymakers, and society in ensuring that AI advancements align with human values and societal well-being. As we navigate this complex intersection of technology and ethics, thoughtful consideration and proactive measures are crucial to shaping an AI-powered future that prioritizes ethical considerations and societal welfare.

Unpredictable Developments

The future of AI presents a landscape of both promise and uncertainty, marked by remarkable progress in specialized tasks. While AI has made significant strides, attaining artificial general intelligence (AGI) remains a formidable and speculative challenge. AGI represents the stage where machines possess humanlike reasoning and adaptability across many tasks, ushering in a new era of technological capabilities. However, realizing this ambitious goal hinges on overcoming complex obstacles, including ethical considerations, computational limitations, and achieving a nuanced understanding of human cognition.

The transformative potential of AI is undeniable, impacting various facets of society, labor markets, and ethical frameworks. As AI continues to evolve, questions surrounding the implications of AGI on employment, privacy, and societal structures become paramount. Ethical considerations, such as bias within AI systems and the responsible use of powerful technologies, add layers of complexity to the trajectory of AI development. Striking a balance between technological advancement and ethical responsibility is crucial to ensuring the positive integration of AI into our lives.

Navigating the unpredictable developments in AI requires a multidisciplinary approach. Researchers, policymakers, and industry leaders must collaborate to address technical challenges and the broader societal impact of AI. Balancing innovation with ethical and societal considerations will be instrumental in steering the trajectory of AI development in a direction that aligns with human values and societal well-being. The journey toward AGI is marked by exploration, challenges, and ethical deliberations, emphasizing the need for a collective effort to shape a future where AI benefits humanity responsibly and ethically.

Nuanced Perspective

Prevalent Dichotomies

Unraveling prevalent dichotomies is crucial in understanding the nuanced impact of AI on various aspects of our lives. Often characterized as a potential threat or a transformative savior, AI's influence is multifaceted and goes beyond binary categorizations. There is a spectrum of implications that challenge oversimplified narratives. Instead of viewing AI in isolation, its dual nature as both a tool and a collaborator in the human experience must be recognized.

By unraveling prevalent dichotomies, it is possible to move toward a more comprehensive understanding of AI's impact, acknowledging its potential benefits and addressing concerns and ethical considerations. This approach will foster a nuanced perspective appreciating the intricate interplay between AI and human dynamics. This will steer discussions toward a more informed and balanced assessment of AI's role in shaping our collective future.

Oversimplified Notions

Challenging oversimplified notions surrounding the AI–human relationship is imperative to foster a more nuanced and comprehensive understanding of this complex interaction. The prevailing tendency to categorize AI dynamics in simplistic terms, such as threats or saviors, overlooks this relationship's intricate layers. Encouraging a more profound understanding involves dismantling these oversimplified notions and acknowledging the multifaceted nature of the AI–human dynamic. It requires embracing the idea that AI is not merely an external force but an evolving collaborator, influencing and being influenced by human behaviors.

In order to encourage a more profound understanding, there is a need for interdisciplinary dialogues that bring together experts from diverse fields, including technology, ethics, sociology, and psychology. This holistic approach will thoroughly explore the ethical, societal, and psychological dimensions of integrating AI into our lives. By challenging oversimplified notions and fostering interdisciplinary discussions, we can pave the way for a richer comprehension of the complexities inherent in the AI–human relationship.

Challenges and Opportunities

Integration of AI into Human Systems

The integration of AI into human systems presents a myriad of challenges that require careful consideration. One significant challenge lies in the ethical implications of AI, as decision-making processes become increasingly automated, raising concerns about bias, transparency, and accountability. The potential displacement of jobs due to automation poses economic challenges, necessitating a thoughtful approach to redefining labor markets and addressing unemployment issues.

Privacy concerns also emerge as AI systems collect and analyze vast amounts of personal data, raising questions about data security and individual autonomy. Additionally, the complexity of AI technologies poses challenges in understanding and interpreting their decisions, creating a gap between the technology and human comprehension.

Opportunities for Coexistence

The integration of AI into human systems brings forth numerous opportunities for responsible and beneficial coexistence. AI technologies can potentially enhance efficiency, productivity, and innovation across various sectors, offering novel solutions to complex problems. Responsible coexistence involves establishing ethical frameworks prioritizing fairness, transparency, and accountability in AI systems. Embracing these opportunities requires collaboration between technologists, policymakers, and society to ensure the responsible development and deployment of AI technologies that align with human values and contribute positively to our collective progress.

Responsible Future

Ethics and the AI–Human Relationship

The ethical implications of the AI–human relationship underscore the critical need for thoughtful consideration as AI continues to integrate into our daily lives. As AI systems become increasingly sophisticated, questions arise concerning issues such as bias, transparency, and accountability in decision-making processes. The potential for AI to impact employment and contribute to social inequality raises ethical concerns that demand careful examination. Privacy becomes paramount as AI technologies handle vast amounts of personal data, prompting discussions on data security and individual autonomy.

Ensuring fairness, inclusivity, and avoiding discrimination in AI algorithms becomes a pivotal ethical challenge. As society grapples with these complex ethical considerations, fostering a dialogue involving technologists, ethicists, policymakers, and the public becomes crucial to establishing ethical frameworks that guide AI's responsible development and deployment, ensuring its alignment with human values and societal well-being.

Ethical Terrain of AI Integration

The rapid advancements in AI bring about unpredictable developments that emphasize the imperative for responsible integration. As AI technologies evolve, unforeseen challenges and implications may arise, demanding a proactive and ethical approach to their incorporation into various aspects of society.

The complexity and sophistication of AI systems can lead to unpredictable outcomes, making it essential to anticipate and address potential risks.

Responsible AI integration requires a careful balance between innovation and ethical considerations, emphasizing transparency, fairness, and accountability in developing and deploying AI technologies. Establishing robust ethical frameworks, implementing regulatory measures, and fostering collaboration among stakeholders is crucial in navigating the unpredictable landscape of AI, ensuring that its integration aligns with human values, respects individual rights, and contributes positively to societal progress.

4

Requirements

AI Requirements Overview

The intricate process of developing and deploying AI systems entails a set of pivotal requirements aimed at ensuring their effectiveness, reliability, and ethical application. This underscores the multidisciplinary nature inherent in AI development, which demands a delicate equilibrium between technical prowess, ethical considerations, user-centric design, and awareness of broader societal impacts. As AI technologies advance, the identification and adherence to these fundamental requirements become increasingly crucial for navigating the complexities of the field. The evolving landscape of AI underscores the need for a comprehensive understanding and integration of these requirements to foster responsible and impactful development.

This chapter serves as an in-depth exploration of the multifaceted dimensions of AI requirements, shedding light on the critical aspects that shape the development and deployment of AI systems. By delving into the intricate details, we aim to provide valuable insights that transcend mere technical considerations. Instead, our exploration encompasses ethical frameworks, user experience principles, and an awareness of the societal implications of AI technologies. This comprehensive approach offers a roadmap for practitioners, developers, and stakeholders, guiding them through artificial intelligence's dynamic and evolving landscape. As we unravel the layers of AI requirements, we strive to equip individuals with the knowledge and tools necessary for responsible and effective engagement in the development and deployment of AI systems.

A. Khan, *Artificial Intelligence: A Guide for Everyone*,
https://doi.org/10.1007/978-3-031-56713-1_4

Navigating the complex terrain of AI requirements requires a nuanced understanding of the ever-changing dynamics in technology, ethics, and societal impact. This chapter not only addresses the current landscape but also anticipates future challenges, providing a forward-looking perspective on the evolving nature of AI requirements. The goal is to empower readers with a holistic view, enabling them to make informed decisions and contributions that align with the ethical and technical standards governing AI development.

Technical Foundations

Data

The foundation of training AI models lies in the availability of high-quality, relevant, diverse, and readily accessible datasets, which form the bedrock of their learning. These datasets should mirror the complexities and nuances of real-world scenarios that the AI system is designed to navigate. The significance of adequate and representative data cannot be overstated, as it forms the raw material upon which machine learning and deep learning algorithms thrive, enabling them to discern intricate patterns, generalize from examples, and ultimately make accurate predictions and informed decisions.

In essence, the quality and comprehensiveness of the data underpin the efficacy and reliability of AI systems, playing a pivotal role in their capacity to address real-world challenges and provide valuable insights across diverse domains. High-quality, relevant, diverse, and more accessible datasets are essential for training AI models. The data should accurately represent the real-world scenarios the AI system will encounter. Adequate data is crucial for machine learning and deep learning algorithms to learn patterns and make accurate predictions.

Algorithms and Models

Selecting or developing suitable algorithms and models is a pivotal step in tailoring AI systems for specific tasks. This decision carries significant weight, as various algorithms are designed to excel in distinct problem domains. The choice of the right algorithm profoundly influences the system's overall performance, efficiency, and effectiveness in addressing the targeted AI task. By aligning the algorithm with the specific problem at hand, it optimizes the system's ability to process data, extract insights, and make accurate predictions, underlining the critical role of algorithm selection in the successful deployment of AI solutions tailored to diverse applications and industries.

The continual development of new and advanced algorithms marks a significant milestone in the evolution of AI systems, facilitating rapid and multifaceted data analysis. These cutting-edge algorithms empower AI systems to process data with remarkable speed and the capacity to examine it at multiple levels concurrently. This, in turn, enables the swift analysis of intricate and multifaceted systems, facilitating predictive capabilities that span a broad spectrum of applications. The synergy between these sophisticated algorithms and AI technology revolutionizes the landscape of data-driven decision-making, enhancing the ability to understand complex systems, forecast future performance, and drive innovation across a diverse range of industries and domains.

Computational Power

AI has witnessed remarkable advancements in recent years with its diverse range of applications, from NLP to computer vision. However, these advancements come hand in hand with substantial computational power requirements. The computational needs of AI systems can be attributed to the complex mathematical computations involved in training and inference processes. Deep learning models, in particular, demand immense computational resources. Training neural networks with millions or even billions of parameters involves processing vast amounts of data through numerous layers, requiring powerful hardware.

Modern GPUs and TPUs have become indispensable tools, enabling the parallel processing essential for training deep learning models efficiently. The computational power required for AI is not solely confined to training. Real-time inference demands substantial processing capabilities, especially in applications like autonomous driving or healthcare diagnostics. As AI continues to evolve, researchers and organizations continually explore innovative hardware solutions, including specialized AI accelerators and distributed computing systems, to meet these growing computational demands.

The computational power requirements of AI extend beyond just the hardware. Efficient software frameworks and algorithms are equally crucial in optimizing resource utilization. Researchers are actively working on developing algorithms that are both computationally efficient and effective, enabling AI systems to perform complex tasks with reduced computational overhead. Additionally, energy efficiency is a significant concern in AI research, as power–hungry systems not only increase operational costs but also have environmental implications. Therefore, there is a growing emphasis on creating AI models and systems that achieve a balance between computational power and energy efficiency, ensuring that AI technology remains sustainable and accessible to a wider range of applications and users.

Expertise

The field of AI is characterized by its interdisciplinary nature, demanding a diverse set of expertise from individuals working within it. At its core, AI necessitates a strong foundation in mathematics, particularly in areas such as linear algebra, calculus, probability theory, and statistics. These mathematical underpinnings are fundamental for understanding and designing AI algorithms. Moreover, programming and computer science expertise is vital for implementing AI solutions effectively. Proficiency in languages like Python, Java, or C++ is essential for coding AI algorithms, building machine learning models, and developing AI applications. Additionally, data manipulation and preprocessing proficiency are crucial, as AI often relies on large datasets for training and inference.

AI expertise extends into specialized domains such as machine learning, natural language processing, computer vision, and reinforcement learning. These subfields require in-depth knowledge and experience in their respective techniques, tools, and libraries. For example, experts in computer vision should be well-versed in image processing techniques, deep learning architectures like Convolutional Neural Networks (CNNs), and relevant libraries such as TensorFlow or PyTorch.

NLP experts should understand the intricacies of language modeling, sentiment analysis, and transformer-based models like BERT. Additionally, expertise in AI ethics and responsible AI practices is becoming increasingly important as the development and deployment of AI systems raise ethical and societal considerations. Consequently, an interdisciplinary collaboration involving experts from various domains is often necessary to tackle complex AI projects effectively.

Validation and Testing

The robustness and reliability of AI systems hinge on a meticulous validation and testing process that rigorously assesses their performance across a spectrum of diverse scenarios. This comprehensive evaluation involves utilizing both real-world and simulated data to thoroughly scrutinize how AI systems operate and identify potential issues or vulnerabilities. Real-world data provides valuable insights into the intricacies of AI systems interacting with the complexities of the actual environment, allowing developers to observe their behavior in authentic conditions. Simulated data, on the other hand, facilitates controlled testing of various conditions and exceptional situations, providing a structured environment to assess the system's responses.

This multifaceted testing approach is pivotal in ensuring that AI systems not only meet their intended objectives but also demonstrate adaptability and resilience when confronted with unforeseen challenges. The significance of such thorough testing and validation cannot be overstated, as it plays a crucial role in the meticulous development and deployment of AI technology, enhancing its overall effectiveness and reliability.

Ethical and User-Centric Requirements

Ethical Considerations

The development of AI systems carries a profound responsibility that necessitates a thorough examination of ethical considerations. Developers are tasked with conscientiously navigating the ethical implications inherent in their work, requiring a commitment to addressing critical issues such as bias present in both data and algorithms, an unwavering dedication to transparency, and strict adherence to ethical guidelines. In AI applications, mitigating bias in data and algorithms stands as a cornerstone for fostering fairness and equity, which should be included in the requirements. Transparency in AI systems plays a pivotal role in ensuring that their operations are not only understandable but also accountable, thereby fostering trust among users and stakeholders.

Furthermore, steadfast adherence to ethical guidelines is paramount to preventing harm to individuals or groups, safeguarding against unintended consequences and ensuring that AI technology is developed and deployed in ways that align with moral principles and societal values. This holistic approach underscores the profound impact of ethical considerations in the AI landscape, emphasizing the commitment to responsible and principled AI development and deployment.

Interpretability and Explainability

Interpretability and explainability are pivotal aspects of AI systems, emphasizing the need for machine learning models to be comprehensible to human beings to an acceptable degree. The demand for clear and intelligible explanations for their decisions intensifies as AI systems become more complex. This is especially crucial in high-stakes domains such as healthcare and finance, where the transparency of AI's decision-making process plays a central role.

The ability to interpret and understand the rationale behind AI-driven decisions is essential for ensuring accountability, fostering trust, and facilitating informed decision-making by human experts. In these critical domains, where the implications of AI decisions can have significant real-world consequences, interpretability and explainability become key requirements for establishing a harmonious collaboration between AI technology and human expertise.

User Experience

User experience is a crucial requirement in AI applications that interact with users, necessitating a focus on delivering a positive and engaging interaction. The cornerstone of achieving this lies in the meticulous design of user interfaces that prioritize intuitiveness and user-friendliness. This involves creating interfaces that enable seamless interactions between individuals and AI systems, promoting a fluid and accessible user journey.

Beyond interface design, ensuring that the behavior of AI systems aligns with user expectations is an essential requirement for cultivating trust and maximizing user satisfaction. This user-centric approach emphasizes not only the technical capabilities of AI but also the incorporation of human-centered design principles. By prioritizing accessibility, comprehensibility, and alignment with the needs and preferences of users, the successful integration of AI into diverse domains is achieved, ensuring a harmonious and positive user experience.

Feedback Loops

Incorporating feedback mechanisms within AI systems is a pivotal aspect of their evolution and enhancement. These mechanisms play a crucial role in enabling AI systems to actively receive input from users and subsequently incorporate that feedback into their decision-making processes. This iterative engagement fosters a continuous cycle of refinement, ultimately leading to improved system accuracy and heightened user satisfaction.

By actively seeking and responding to user insights, AI systems not only refine their performance but also cultivate trust and user confidence. This dynamic relationship between technology and users drives the iterative development of AI systems to better meet the evolving needs and expectations of their user base. The integration of feedback loops thus emerges as an integral strategy for ensuring that AI systems remain adaptable, responsive, and attuned to the dynamic landscape of user preferences and requirements.

System Dynamics and Governance

Scalability

Scalability stands as a fundamental aspect in the intricate landscape of AI system design, representing the system's inherent ability to accommodate an array of scales concerning data and tasks flexibly. The hallmark of a well-designed AI system lies in its adaptability, showcasing a core attribute that is essential for its seamless integration into diverse and dynamic environments. One of the pivotal challenges that AI systems encounter is the management of expanding workloads, which may manifest in the form of growing datasets or increasingly complex tasks.

The robustness of these systems is a critical requirement in ensuring that they can navigate through such challenges without significant compromises in performance quality. This adaptability not only bolsters the reliability of AI systems but also enhances their relevance across various applications and industries. As technology continues to evolve, the scalability of AI becomes an increasingly crucial requirement, positioning these systems as versatile tools capable of meeting the demands of an ever-changing landscape.

Security and Privacy

Security and privacy are paramount requirements of AI systems, particularly given their frequent interaction with sensitive data. The onus lies on developers to institute robust security measures that extend beyond the AI system to encompass the entirety of the data it accesses and processes. These measures serve a dual purpose: safeguarding against external threats and upholding the principles of data integrity, confidentiality, and privacy compliance.

Developers play a crucial role in embedding security and privacy as integral components of AI development, thereby mitigating potential risks and reinforcing trust among users. This approach not only aligns with ethical and legal standards but also ensures responsible AI deployment, where the foundational principles of security and privacy are prioritized for the benefit of users and the broader ethical framework.

Regulations and Compliance

AI systems must adhere to all regulations in the environment in which they operate. Therefore, any regulations and compliance requirements must be incorporated into the AI system specifications. Navigating the regulatory landscape for AI systems is complex, given the significant variations based on the application domain and jurisdiction. Developers and stakeholders are responsible for remaining vigilant in understanding and complying with the relevant regulations governing their specific context. These regulations span critical aspects, including but not limited to data protection, safety standards, and the equitable treatment of individuals or groups. Recognizing that adherence to this intricate regulatory framework goes beyond mere compliance ensures that AI systems are legally sound and align with societal expectations.

By designing, developing, deploying, and utilizing AI in a manner that respects legal mandates and societal norms, stakeholders contribute to fostering trust, accountability, and responsible AI practices that transcend geographical boundaries and diverse contexts. This approach underscores the commitment to the ethical and lawful deployment of AI technology, instilling confidence in users and maintaining the integrity of AI applications worldwide.

Continuous Learning and Adaptation

The design of AI systems must encompass the capacity to continually learn from fresh data and adapt to evolving circumstances as they unfold. To facilitate this dynamic capability, developers should integrate mechanisms for ongoing learning and the seamless updating of AI models. This iterative learning process ensures that AI systems remain relevant and effective in an ever-changing environment, enabling them to incorporate novel insights and respond adeptly to shifting patterns and complexities. By embracing this iterative approach, AI systems evolve in tandem with the challenges and opportunities presented by their respective domains, underlining their adaptability and enduring relevance in an era of rapid technological advancement.

Sustainability

The growing complexity and resource requirements of AI systems have brought heightened attention to their energy consumption and environmental impact, emphasizing the imperative for sustainable AI development practices, which need to be incorporated into the requirements. To address these considerations, a strategic focus on optimizing the energy efficiency of AI systems has become paramount, especially in their training and inference phases. This optimization aims to minimize the carbon footprint associated with AI technologies and reduce overall energy consumption.

Additionally, adopting renewable energy sources and utilizing eco-friendly hardware configurations serve as crucial steps in mitigating the environmental impact of resource-intensive AI applications. In the current era of rapid AI advancement, responsible integration of these technologies demands a conscientious approach that harmonizes technical progress with environmental sustainability. This approach seeks to strike a balance, ensuring that AI systems not only deliver powerful capabilities but also adhere to principles of environmental responsibility, thereby contributing to a sustainable and eco-friendly technological landscape.

5

Technologies, Techniques, and Components

Overview

AI Components

AI is an umbrella term that includes any type of software or hardware component that supports machine learning, computer vision, natural language understanding (NLU), and NLP, as well as other technologies and techniques that contribute to AI.

AI is a multifaceted domain that defies a singular definition, instead encompassing a wide spectrum of technologies, approaches, and techniques geared toward enabling machines to replicate or simulate human intelligence and behaviors. This expansive field extends its reach into a variety of disciplines, weaving together the threads of computer science, mathematics, neuroscience, and more. This interdisciplinary nature underscores AI's versatility, allowing it to embrace an array of methods, from rule-based systems and machine learning to NLP and computer vision, all aimed at imbuing machines with cognitive capabilities reminiscent of human thought and action.

Functionality of AI Components

The integration and collaboration of various technologies and techniques underpin the functionality of AI components. These elements can operate synergistically or independently, collectively contributing to the creation of intelligent systems that significantly impact our daily lives and various

© The Author(s), under exclusive license to Springer Nature Switzerland AG 2024
A. Khan, *Artificial Intelligence: A Guide for Everyone*,
https://doi.org/10.1007/978-3-031-56713-1_5

industries. In comprehending the multifaceted nature of AI, it becomes evident that its development frequently necessitates the convergence of diverse approaches to attain specific desired outcomes.

As a field, AI is characterized by its continuous evolution, marked by the emergence of novel technologies and techniques that are constantly integrated to address intricate challenges. This dynamism reflects the adaptability of AI as researchers and developers strive to enhance its capabilities and address new frontiers. The amalgamation of machine learning, natural language processing, computer vision, and other AI subfields exemplifies the interdisciplinary nature of AI development. Within this amalgamation, AI's true potential unfolds, harnessing the strengths of different approaches to create intelligent systems with a broad spectrum of applications.

The collaborative interplay of AI components is pivotal in realizing advancements across diverse domains. Machine learning algorithms, for instance, can analyze vast datasets to identify patterns and make predictions, while natural language processing enables systems to understand and generate human-like language. On the other hand, computer vision empowers AI systems to interpret and make decisions based on visual information. The orchestration of these diverse components not only amplifies the capabilities of AI but also allows for the creation of more sophisticated and adaptable intelligent systems.

Moreover, the dynamic landscape of AI development emphasizes the need for continuous learning and adaptation. As new challenges emerge and societal demands evolve, AI researchers and practitioners explore innovative technologies and refine existing techniques to address contemporary issues. This iterative process of improvement and integration ensures that AI remains at the forefront of innovation, consistently pushing the boundaries of what is achievable.

In summary, the workings of AI components embody a dynamic and interdisciplinary landscape where the amalgamation of technologies and techniques creates intelligent systems that impact diverse facets of our lives and industries. The continuous evolution of AI underscores the importance of staying abreast of emerging technologies, fostering an environment where innovation thrives, and intelligent systems continually advance to meet the demands of our ever-changing world.

Technologies and Techniques

Machine Learning

Machine learning is a subset of AI that involves training computers to learn from data and improve their performance over time. This includes tasks like image recognition, language translation, and self-driving cars. Machine learning algorithms enable computers to recognize patterns and make predictions without being explicitly programmed for each specific case. It includes various algorithms like decision trees, support vector machines, and neural networks.

Deep Learning

Deep learning is a type of machine learning that uses neural networks inspired by the structure of the human brain. It uses artificial neural networks with many layers to model and process complex patterns in data. It is particularly powerful for tasks like image and speech recognition. Deep learning has enabled advancements in areas like computer vision and natural language processing.

Natural Language Processing

Natural language processing focuses on enabling computers to understand, interpret, and generate human language. It powers chatbots, language translation, sentiment analysis, and text generation. This technology is described in detail in a subsequent chapter.

Computer Vision

AI boasts impressive image-related capabilities, excelling in image recognition as it can discern and label objects, people, and scenes within both images and videos. It showcases object-detection proficiency, pinpointing and labeling specific objects within visual content, and enhancing applications like autonomous vehicles and security systems.

AI's facial recognition prowess enables it to identify individuals based on facial features, which finds utility in security, access control, and personal devices. Also, AI's applications extend into the realm of healthcare, where it plays a crucial role in medical image analysis, aiding in disease diagnosis by scrutinizing medical images such as X-rays and MRIs, thereby enhancing the accuracy and efficiency of healthcare diagnostics.

Robotics

The fusion of AI and mechanical engineering represents a transformative synergy that yields machines capable of autonomous or semiautonomous operation, requiring minimal human intervention. These AI-driven mechanical systems find application in diverse fields, ranging from manufacturing and healthcare to exploration and even household chores. In manufacturing, autonomous robotic systems streamline production processes, optimizing efficiency and precision.

In healthcare, AI-augmented devices assist medical professionals in diagnostics and surgeries, enhancing patient care. In exploration, autonomous drones and vehicles navigate challenging terrains and gather data in remote locations. Even in everyday life, AI-driven household robots undertake tasks such as vacuuming and mowing lawns, exemplifying the profound impact of this convergence of automation and human convenience across various domains.

Expert Systems

Expert systems, a category of rule-based AI programs, are meticulously crafted to replicate the decision-making process of human experts within particular domains of expertise. These systems have found prominent roles in critical sectors, including medicine and finance, where their ability to diagnose diseases or provide investment recommendations is invaluable. In the realm of healthcare, expert systems assist in the diagnostic process, leveraging extensive rule sets to analyze patient data and offer medical insights.

Similarly, these AI systems harness their rule-based logic in finance to assess market trends and deliver investment guidance. By amalgamating human expertise with AI's computational precision, expert systems contribute to informed decision-making in specialized domains, highlighting their pivotal role in augmenting human capabilities and driving efficiency.

Reinforcement Learning

Reinforcement learning constitutes a fundamental paradigm of AI training, characterized by the iterative process through which AI agents acquire expertise. Within this framework, AI agents learn through a continuous cycle of trial and error, interacting with their environment and making decisions based on their current knowledge. Crucially, these decisions yield consequences, and the AI agent receives feedback in the form of rewards or penalties, allowing it to refine its decision-making strategies over time.

This dynamic learning process imbues AI agents with the capacity to navigate complex, uncertain environments, making reinforcement learning a pivotal approach for imbuing AI with adaptability and proficiency in scenarios where explicit programming is impractical.

Genetic Algorithms

Inspired by natural selection principles, genetic algorithms serve as a powerful optimization technique employed to unearth solutions to intricate problems. This innovative approach operates by simulating the process of evolution, where a population of potential solutions undergoes iterative refinement. These solutions are treated as individuals within a population and are subjected to genetic operations such as selection, crossover, and mutation, mirroring the mechanisms of genetic variation and selection observed in nature.

Through successive generations, genetic algorithms hone in on optimal or near-optimal solutions, making them a versatile tool in tackling complex, multifaceted challenges across various domains, from engineering and finance to artificial intelligence and beyond.

Fuzzy Logic

Fuzzy logic, a remarkable concept in AI, equips systems with the capacity to grapple with imprecise or uncertain data, effectively mirroring the nuanced and flexible nature of human reasoning. Unlike conventional binary logic, which strictly adheres to true or false outcomes, fuzzy logic accommodates situations marked by varying degrees of truth or ambiguity. This inherent adaptability allows AI systems to navigate the complexities of real-world scenarios where decisions often dwell in the realm of "shades of gray."

By embracing fuzzy logic, AI systems demonstrate an ability to make contextually informed decisions that resonate with the intricacies of human cognition, facilitating their application in domains where precise, black-and-white answers are elusive and where more nuanced, real-world problem-solving is imperative.

Neuromorphic Computing

Neuromorphic computing represents a cutting-edge approach in AI that seeks to replicate the architecture and functionality of the human brain within hardware. This innovative emulation of the brain's neural structure facilitates AI systems in processing information with remarkable efficiency and enables them to mimic human-like learning processes. By modeling AI systems on the principles of the brain's neural networks, neuromorphic computing not only enhances their capacity for complex, context-aware decision-making but also accelerates advancements in areas like machine learning, pattern recognition, and cognitive computing.

This convergence of hardware and neuroscience promises to unlock new frontiers in AI, allowing technology to evolve closer to human cognition and fostering applications that excel in tasks requiring adaptability, inference, and learning from experience.

Cognitive Computing

Cognitive computing represents an ambitious endeavor within the realm of artificial intelligence, aspiring to replicate human thought processes within computerized models. By harnessing this technology, machines gain the ability to interact with humans in ways that are notably natural and intuitive. This entails the integration of various AI techniques, including natural language processing, machine learning, and pattern recognition, to enable AI systems to comprehend, reason, and respond to human inputs with a depth and nuance that resembles human cognitive capabilities.

The ultimate goal of cognitive computing is to bridge the gap between human and machine intelligence, paving the way for AI systems that are more perceptive and context-aware and capable of delivering more enriching and human-like interactions across a spectrum of applications and domains.

AI Frameworks

AI frameworks, comprising a suite of tools and software libraries, serve as the foundational infrastructure empowering developers in the creation of AI models and applications. These indispensable resources streamline the development process, alleviating complexities and expediting the integration of AI technologies into diverse applications. By offering a cohesive ecosystem that encompasses machine learning algorithms, data-handling capabilities, and optimization techniques, AI frameworks significantly simplify the intricate task of harnessing the potential of artificial intelligence.

Developers can leverage these frameworks to accelerate their projects, enabling the rapid deployment of AI solutions ranging from natural language processing and computer vision to predictive analytics and beyond, fostering innovation and accessibility within the AI landscape.

Autonomous Systems

Autonomous systems, exemplified by self-driving cars and drones, represent a paradigm shift in technology, harnessing the power of AI to operate independently and make decisions without the need for human intervention. These cutting-edge innovations are equipped with a sophisticated array of sensors, cameras, and AI algorithms, which collectively enable them to perceive and interpret their environment in real-time. This perceptual prowess enables autonomous systems to navigate through complex and dynamic surroundings, make informed decisions, and execute actions autonomously.

Whether it is a self-driving car safely navigating city streets or a drone conducting precise aerial maneuvers, these AI-driven autonomous systems are poised to revolutionize industries ranging from transportation and logistics to agriculture and beyond, reshaping the way we interact with technology and the world around us.

Virtual Assistants

As exemplified by Apple's Siri, Google Assistant, and Microsoft's Cortana, virtual assistants are sophisticated AI-driven software applications that seamlessly engage with users through text or speech-based interactions. These intelligent digital companions are adept at answering questions, performing various tasks, and furnishing users with valuable information. Leveraging natural language processing and machine learning capabilities, virtual assistants offer an intuitive and user-friendly experience catering to a diverse range of user needs.

From setting reminders and retrieving weather updates to initiating web searches and controlling smart home devices, these AI-driven virtual assistants have become integral components of our daily lives, offering a glimpse into the transformative potential of AI technology in enhancing user convenience and productivity.

Gaming AI

In the realm of video games, AI plays a pivotal role by infusing non-player characters (NPCs) with human-like behavior and adaptability, thereby elevating the gaming experience to new heights. AI algorithms empower these NPCs to react dynamically to player actions, ensuring that the gameplay remains engaging and challenging. By mimicking human decision-making and learning, AI-driven NPCs enhance immersion and interactivity, creating game worlds where opponents and allies exhibit a level of intelligence that rivals human players. This not only enriches the overall gaming experience but also underscores the profound impact of AI in revolutionizing the entertainment industry, ushering in an era where virtual worlds are populated by AI entities that blur the line between human and artificial intelligence.

High-Level Components

At a high level, the three key components required by AI to achieve its objectives, though all are not specific to AI, are as follows:

- Computational systems
- Data and data management
- Advanced AI algorithms (code)

Computational Systems

The first crucial component for the functioning of AI, computational systems, provide the necessary processing power to execute complex algorithms and models, enabling AI applications to analyze data, recognize patterns, and make informed decisions. The computational infrastructure is the backbone of AI, supporting the intense calculations and computations required for various tasks. The more human-like the desired outcome, the more data and processing power are required.

Data and Data Management

The second essential component is data and data management. Data serves as the lifeblood of AI, providing the information necessary for training algorithms and making predictions. Effective data management involves collecting, organizing, and maintaining datasets to ensure the quality and relevance of information fed into AI systems. High-quality, diverse datasets are fundamental for training AI models, allowing them to generalize and perform well across different scenarios.

Advanced AI Algorithms (Code)

The third key component for AI success is advanced AI algorithms, often represented in code. These algorithms are the set of instructions that guide the AI system's decision-making process. From machine learning models to deep neural networks, the sophistication of these algorithms determines the AI system's capability to learn, adapt, and perform complex tasks. Crafting efficient and effective algorithms is a crucial aspect of AI development, as they define how AI interprets data and produces outcomes.

Detailed-Level Components

At a more detailed level, AI systems are composed of various components that work together to enable intelligent behavior, learning, and decision-making capabilities in AI systems. These components can vary depending on the specific AI application and approach. The choice of components depends on the specific AI task and the goals of the applications. The common components required by AI systems are described in the following sections.

Data

The significance of high-quality and pertinent data cannot be overstated when it comes to AI, as it forms the bedrock upon which AI models are constructed and refined. This multifaceted data serves as the raw material for training and testing AI algorithms, encompassing a wide spectrum of information types, from textual content and images to audio recordings, video streams, sensor readings, and more.

The choice and quality of data are paramount, as they profoundly influence the performance and robustness of AI systems across diverse applications. By leveraging this diverse data landscape, AI models gain the capacity to excel in tasks that span natural language understanding, computer vision, voice recognition, predictive analytics, and more, exemplifying the pivotal role that data quality and diversity play in advancing the frontiers of artificial intelligence.

Input Interface

The sensory input component represents a critical facet of AI systems, serving as their interface to the external world. Its versatility is showcased through its ability to accept a wide array of data formats, including text, voice, images, and various other data types, contingent on the specific application. This component acts as the gateway through which AI systems ingest information from their environment, enabling them to perceive and interact with the world in a manner analogous to human sensory perception.

Whether it is processing spoken language, analyzing visual scenes, or handling diverse data streams, this sensory input interface underpins AI systems' capacity to comprehend and respond to the rich tapestry of data that characterizes the real world, contributing to their adaptability and utility across a myriad of domains and contexts.

Preprocessing

Data preprocessing stands as a pivotal stage in the data analytics pipeline, encompassing the crucial tasks of cleansing, transforming, and formatting raw data to render it amenable for subsequent analysis and model training. This indispensable process is undertaken with the dual objectives of enhancing data quality and mitigating noise, thereby fortifying the robustness and reliability of downstream data-driven endeavors.

By meticulously curating datasets, handling missing values, standardizing data formats, and performing relevant transformations, data preprocessing empowers data scientists and AI practitioners to glean valuable insights and construct accurate predictive models. It serves as the initial crucible where raw data is refined into a coherent, structured foundation that fuels the generation of meaningful and actionable knowledge, illustrating its fundamental role in the data-driven decision-making landscape.

Feature Extraction

Feature extraction occupies a pivotal position in the data preprocessing continuum, entailing the meticulous process of either handpicking or generating pertinent features from the dataset, which subsequently serve as the input variables for AI models. These features play an indispensable role in encapsulating salient patterns, inherent structures, and distinctive characteristics intrinsic to the data. By judiciously curating the selection or engineering of features, data scientists and machine learning practitioners harness the power to unveil meaningful insights, enhance model interpretability, and bolster predictive accuracy.

Feature extraction serves as the linchpin in the transformation of raw data into a refined, structured representation, empowering AI models to effectively distill and capitalize on the intricate nuances and relevant information embedded within the dataset.

Machine Learning Algorithms

Tailoring the choice of machine learning algorithms to the specific requirements of the AI task at hand is a critical decision in the development of intelligent systems. The diverse landscape of machine learning techniques offers various options, each suited to particular problem domains and data characteristics. Decision trees excel in transparent decision-making, support vector machines find utility in classification tasks, k-nearest neighbors (k-NN) thrive in proximity-based scenarios, and neural networks demonstrate prowess in complex pattern recognition.

The judicious selection of these algorithms, informed by a deep understanding of the task's intricacies and data properties, empowers data scientists and machine learning practitioners to construct models that align seamlessly with the objectives, effectively extracting insights and making predictions across domains as varied as finance, healthcare, and natural language processing, among others.

Training Data

The training of AI models unfolds through the utilization of either labeled data, characterizing supervised learning, or unlabeled data, which typifies unsupervised learning paradigms. In both scenarios, this training data serves as the foundational bedrock upon which AI models are cultivated, enabling them to grasp intricate patterns, discern relationships, and infer valuable insights from the information contained within.

Supervised learning, underpinned by labeled data, involves the provision of explicit guidance in the form of predefined target outcomes, facilitating the model's ability to map input features to desired outputs. Conversely, unsupervised learning navigates the uncharted waters of unlabeled data, tasking the model with the autonomous discovery of latent structures, clustering, or data representations. This fundamental phase of model training represents the crucible where AI systems are imbued with the knowledge and acumen essential for subsequent data-driven decision-making and prediction tasks, highlighting its central role in the landscape of artificial intelligence.

Model Architecture

Within the domain of deep learning, the architecture of a model serves as the blueprint dictating the intricate structure of the neural network. This architectural framework encompasses crucial elements such as the number of layers, the arrangement of nodes within each layer, and the intricate web of connections that interlink them. Selecting an appropriate architecture is a pivotal decision, as it profoundly influences the model's capacity to excel in specific tasks.

Diverse architectural designs, ranging from convolutional neural networks for image analysis to recurrent neural networks (RNNs) for sequential data, cater to a vast spectrum of AI applications. The intricacies of the chosen architecture shape the model's ability to extract hierarchical features, capture nuanced patterns, and effectively address the unique challenges posed by the task at hand, underlining the pivotal role of architectural design in the domain of deep learning.

Learning Algorithm

The learning algorithm stands as a critical pillar in the training process of AI models, governing the manner in which the model iteratively adjusts its internal parameters in response to the training data it encounters. A prevalent and foundational algorithm in the field of neural networks is backpropagation, which orchestrates the model's learning by propagating error signals backward through the network, allowing it to fine-tune its parameters.

This iterative optimization process is paramount in enabling the model to refine its predictive capabilities, aligning them with the intricacies and nuances embedded within the training data. The choice and implementation of the learning algorithm wields a profound impact on the model's convergence, speed of learning, and generalization performance, highlighting its pivotal role in developing and refining AI systems.

Feature Transformation

In certain scenarios, data preprocessing strategies encompass feature transformation techniques, among which dimensionality reduction methods like principal component analysis (PCA) emerge as prominent tools. These techniques serve the dual purpose of simplifying complex datasets while preserving the salient and informative aspects of the original data. By distilling the data into a reduced set of dimensions, feature transformation mitigates issues stemming from high dimensionality and enhances computational efficiency and model interpretability.

PCA, for example, helps find important patterns in data by identifying the most significant directions of variation. This allows data scientists to simplify their analysis while retaining essential patterns and relationships. These methods are useful tools that make it easier to process and model data in various AI applications.

Inference Engine

The inference component of an AI system represents the culmination of its training phase, stepping into action when confronted with fresh, uncharted data. Its primary function lies in the judicious application of the knowledge, patterns, and insights acquired during the model's training to make informed predictions or decisions on these novel instances.

Whether it entails classifying an image, transcribing spoken language, or forecasting future trends, the inference component serves as the AI system's bridge to real-world applicability. By applying the learned patterns to unexplored data, it unlocks the power of AI for tasks ranging from natural language understanding to computer vision, transforming raw data into valuable insights and actionable outcomes across diverse domains and use cases.

Output Interface

The output interface of AI systems assumes a pivotal role in presenting the outcomes of AI processing, rendering them accessible and comprehensible to both human and machine recipients. It acts as the conduit through which AI analysis and decision-making results are conveyed, manifesting in a myriad of formats encompassing text, visuals, audio, or even tangible actions orchestrated by physical systems. This interface enhances the interpretability and usability of AI-generated insights and enables seamless integration with downstream processes and user interactions.

Whether it is a textual summary of a financial report, a visual representation of data trends, or the execution of commands by an autonomous robot, the output interface transforms AI-generated output into a meaningful and actionable form, facilitating its utility across a diverse array of applications and domains.

Feedback Loop

The establishment of a feedback loop is of paramount importance for AI systems characterized by their capacity to learn and adapt iteratively. This essential mechanism serves as the conduit through which valuable input and insights are collected from users or the environment, subsequently channeling this feedback to enhance and fine-tune the AI's performance. By assimilating real-world experiences, user preferences, and evolving data, the AI system embarks on a continuous journey of self-improvement, refining its predictions, decision-making, and overall efficacy.

This dynamic feedback loop not only bolsters the AI system's adaptability but also fortifies its capacity to align more closely with its users' ever-changing requirements and expectations and the broader context in which it operates, thereby underlining its indispensable role in the evolution of AI technologies.

Optimization Techniques

The optimization of AI models is a pivotal undertaking aimed at fine-tuning the model's parameters to achieve optimal performance characterized by minimized errors or maximized accuracy. This optimization process, akin to navigating a multidimensional landscape in search of the lowest valleys, involves applying specialized techniques, with gradient descent emerging as a prominent method. Gradient descent is like fine-tuning a model to make it better. It does this by adjusting the model's settings based on how far off its predictions are from the actual results. The goal is to find the best settings to make the model work well.

Through this iterative and data-driven process, AI practitioners harness optimization methods to unlock the full potential of their models, enhancing their predictive capabilities and ensuring they align closely with the task's objectives, whether it pertains to image classification, natural language processing, or any other AI application domain.

Evaluation Metrics

The measurement of AI model performance hinges on the deployment of evaluation metrics, which act as quantitative yardsticks to gauge the model's effectiveness in various tasks. An array of metrics exists for this purpose, encompassing standard measurements like accuracy, precision, recall, and F1-score, among others. Accuracy reflects the proportion of correct predictions made by the model, while precision quantifies the proportion of true positive predictions out of all positive predictions, and recall measures the ability to capture all true, positive cases.

The F1 score harmonizes precision and recall into a single metric, offering a balanced view of the model's performance. These evaluation metrics serve as vital instruments in the assessment of AI models, facilitating data-driven decision-making and continual refinement to ensure the models align closely with the desired outcomes across diverse application domains, from healthcare to image recognition.

Deployment Infrastructure

The deployment of AI models represents a pivotal transition from the training phase to practical, real-world application, necessitating the establishment of a robust infrastructure capable of hosting, serving, and managing these models efficiently. This infrastructure is often constructed leveraging cloud services, offering scalability, reliability, and accessibility. This deployment makes AI models accessible to users and systems, enabling them to interact with and benefit from the model's predictive or decision-making capabilities.

Whether it involves deploying a chatbot on a website, an image recognition system in a mobile app, or a recommendation engine in an e-commerce platform, this phase bridges the gap between AI development and tangible utility, making AI-powered solutions accessible to a diverse range of industries and applications.

Ethical Considerations

Responsible AI encompasses a multifaceted approach that encompasses critical components designed to uphold ethical and fair behavior within AI systems. These components include robust mechanisms for detecting and mitigating bias in data and algorithms, thereby minimizing the risk of perpetuating unfair or discriminatory outcomes. Additionally, fairness assessment frameworks are integral in evaluating AI system outputs to ensure equitable treatment across different demographic groups and contexts.

Transparency initiatives play a crucial role by shedding light on the decision-making processes within AI systems, making them more understandable and interpretable. By integrating these elements, responsible AI endeavors to foster ethical, unbiased, and transparent AI systems, thereby fostering trust and accountability in an increasingly AI-driven world across various domains, including finance, healthcare, and justice.

Monitoring and Maintenance

AI systems require continuous monitoring and maintenance to safeguard their sustained performance, adaptability, and resilience against degradation. This essential process involves vigilant oversight, encompassing aspects such as data quality assurance, model performance tracking, and proactive error identification and correction. By consistently assessing the system's outputs and comparing them to expected results, AI practitioners can swiftly identify and rectify any deviations or deteriorations in performance.

Also, as the data landscape evolves and user requirements shift, the AI system's algorithms and models may need periodic updates and fine-tuning to remain effective and aligned with the ever-changing, real-world context. This ongoing vigilance and maintenance ensure that AI systems not only meet their intended objectives but also continue to deliver valuable insights and predictions across diverse domains and applications.

6

Building an AI System

Crafting AI: Building Intelligent Systems

Process

AI systems operate through the synergy of extensive datasets and iterative processing algorithms, a harmonious combination that enables them to glean insights and knowledge from patterns and features within the data. By scrutinizing these datasets, AI systems harness the power of algorithms to execute tasks that traditionally demand human intelligence.

The operation of an AI system unfolds in a series of distinct steps, a sequence that can diverge depending on the AI system's type and its designated task. At each juncture of data processing, the AI system methodically processes information, evaluating its performance as it proceeds. The results of this internal assessment are invaluable, serving as building blocks for the system to accumulate knowledge and refine its expertise. In essence, the AI system learns through a continuous cycle of processing, self-assessment, and adaptation.

Reverse Engineering Human Traits

In its most basic form, computers are programmed to mimic human behavior using data from similar past behavior. This can range from recognizing differences between a car and a truck to performing complex activities in a manufacturing facility. In simple terms, this means making computers think and act like humans.

© The Author(s), under exclusive license to Springer Nature Switzerland AG 2024
A. Khan, *Artificial Intelligence: A Guide for Everyone*,
https://doi.org/10.1007/978-3-031-56713-1_6

The process of building an AI system involves reverse-engineering human traits and capabilities in a machine and then using its computational capabilities to surpass what humans are capable of. In practical terms, this involves several steps and considerations, from defining the problem to deploying the final solution.

Leveraging Computer Processing Power

The complexity of AI systems can vary widely based on the task, the algorithms used, and the data available. Some AI systems are rule-based and deterministic, while others, like deep learning models, are based on complex statistical computations and require large volumes of data for effective learning. Overall, AI systems strive to emulate human intelligence and decision-making processes while leveraging the power of computational processing.

Process Iteration

The development of AI systems is inherently dynamic, marked by a perpetual cycle of collaboration, learning, and adaptability. It is vital to understand that the creation of AI solutions isn't a straightforward, linear path; rather, it typically mirrors an iterative and ever-evolving process. In this fluid journey, collaboration between multidisciplinary teams, including data scientists, domain experts, and software engineers, plays a central role in fostering innovation and effectively addressing the complexities of real-world challenges. This iterative and adaptable approach allows AI systems to continually refine their capabilities and maintain their relevance as they interact with evolving data and user requirements.

AI Learning

Through progressive learning algorithms, AI adapts to let data do the programming. It finds structure and patterns in data so that the algorithms can learn. For example, an algorithm can teach itself how to play chess, pick stocks, or recommend products online. As new data is added, it adapts based on the newly acquired data.

AI can analyze more and deeper data using neural networks with many processing layers. Due to intensive computing requirements, applications requiring deep processing layers, such as those required for credit card fraud detection, were previously impossible to build. Now, using AI with its powerful GPUs and models makes such capabilities easy to implement because the models learn directly from the data.

AI can also help achieve accuracy through neural networks. For example, Siri and Alexa, based on deep learning, keep getting more accurate as we use them.

AI Development Steps

The AI development process typically involves several stages or steps. While the specific details may vary depending on the project and the application, there are five basic steps commonly followed in AI development, which are explained in the following subsections.

Step 1: Problem Definition and Planning

Problem Definition

In embarking on an AI-driven solution, it is essential to define the specific problem that needs to be addressed. This initial step involves a thorough understanding of the problem's nuances and establishing clear objectives, constraints, and desired outcomes. It is crucial to assess whether AI is the appropriate tool for solving the problem, considering its suitability and effectiveness in achieving the desired goals and weighing it against other potential solutions. This thoughtful and methodical approach ensures that AI is leveraged optimally to tackle the problem at hand.

Data Collection

The next step in preparing for the AI solution involves identifying the data sources and collecting the necessary data for training, testing, and validating the AI model. Data may originate from a variety of sources, including sensors, databases, or external application programming interfaces (APIs). It is imperative to ensure that the data is both relevant and of high quality to ensure the AI model's accuracy and effectiveness.

The input data can be in the form of text, images, audio, or any other data type, depending on the task. This data serves as the information on which the

AI system will make decisions or predictions. Proper data collection and management form the bedrock of any successful AI application, as the model's performance is intricately tied to the quality and appropriateness of the data used.

Project Scope

In the planning phase of the AI project, it is crucial to define the project's scope comprehensively, outlining the necessary resources, timeline, and any potential constraints. This step allows for a clear understanding of what it will take to implement the AI solution successfully. By establishing the scope, it becomes possible to effectively allocate resources, set realistic timelines, and identify any potential limitations that must be addressed to ensure the project's success.

Step 2: Data Preprocessing and Preparation

Data Preprocessing

The data cleaning and preprocessing phase is vital to ensure the data's quality and suitability for AI model training. During this step, noise, missing values, and other data quality issues are addressed, enhancing the data's reliability. Additionally, data is formatted appropriately to make it compatible with the chosen AI model, setting the stage for accurate and effective training. This meticulous process of data preparation is essential for the model's success, as it helps eliminate potential sources of errors and inaccuracies in the AI system's predictions and decisions.

Feature Engineering

Features are specific attributes or characteristics that the AI model will use to make decisions. For example, features could be edges, colors, or shapes in an image recognition task. Feature selection and engineering are crucial components of the AI model development process. If necessary, the AI system extracts relevant features from the input data. In this phase, the developer either chooses relevant features from the dataset or creates new ones that will serve as inputs for the AI model.

The choice of features directly impacts the model's performance and its ability to make accurate predictions, making this step a critical element in the model's overall effectiveness. The model can be fine-tuned through thoughtful feature engineering to capture and utilize the most pertinent information, thus enhancing its predictive capabilities.

Data Splitting

Data is typically partitioned into training, validation, and test sets to develop a robust AI model. The training set is used to teach the model, the validation set helps fine-tune its parameters, and the test set is reserved for evaluating its performance, ensuring that the model generalizes well to new, unseen data. This division enables the iterative refinement of the model and helps gauge its predictive accuracy.

Step 3: Model Development and Training

Model Selection

Selecting the right machine learning or deep learning algorithm is a pivotal decision in the AI model development process. The appropriate algorithm or model is chosen based on the task. For instance, image recognition might use convolutional neural networks, while language translation could involve sequence-to-sequence models.

The choice of algorithm should be tailored to the problem at hand and the characteristics of the data being used. Factors such as the nature of the data, its dimensionality, and the complexity of the problem play a significant role in determining whether a specific algorithm is suitable. By making a well-informed choice in algorithm selection, developers can optimize the model's performance and ensure that it aligns with the specific demands of the task.

Model Training

Training the selected model is a pivotal stage in AI development, involving the utilization of the training data to enable the AI system to learn from the provided information. In this phase, the AI model acquires knowledge from labeled data, where the outcomes are known.

Hyperparameters, which are settings that influence the model's learning process but are not derived from the data, play a significant role. They are meticulously fine-tuned and optimized to enhance the model's performance. These hyperparameters govern the learning process, ensuring that the AI model converges effectively.

A model uses optimization techniques to adjust its internal parameters, including weights and biases (which are akin to extra adjustments) in neural networks, based on the training data. By presenting the input data along with their corresponding correct outputs, the model strives to minimize errors and make accurate predictions. This iterative process is the cornerstone of AI model training.

Techniques like grid search or random search can be employed to systematically explore various hyperparameter configurations and achieve a more effective AI model. These methods enable the model to refine its settings and improve its ability to make precise predictions.

Finally, the AI system produces output based on the inference process. The output could be a classification label, a translated sentence, a recommendation, or any other form of decision or action.

Validation and Testing

After training, the AI model is ready for inference. During inference, the model processes new, unseen data to make predictions or decisions. This is the phase where the AI system demonstrates its learned intelligence.

Separate validation and testing datasets are used to evaluate the AI model's effectiveness and its ability to make accurate predictions on unseen data. The validation set allows for refining the model's parameters while the testing dataset assesses its generalization performance, ensuring that it performs well on new, unobserved data, which is a crucial aspect of AI model validation.

Step 4: Model Evaluation and Testing

Performance Evaluation

Evaluating the AI model's performance is a critical step in ensuring its capability to effectively generalize unseen data. This evaluation entails the use of appropriate metrics to gauge the model's effectiveness, which can include standard measures like accuracy, precision, recall, and F1-score. Additionally, domain-specific metrics may be applied when the nature of the problem demands a specialized evaluation approach. These performance assessments are fundamental in determining the model's reliability, suitability, and overall success in solving the targeted problem.

Iterative Improvement

Upon reviewing the evaluation results, the model development process often enters an iterative phase where adjustments are made to improve the model's performance. This includes iterative cycles of training, evaluation, and improvement. Fine-tuning the model, tweaking its features, or considering alternative algorithms are common strategies employed to address performance issues or enhance its predictive capabilities. This iterative approach continues until the model attains the desired level of performance, ensuring that it effectively solves the problem at hand and meets the predefined objectives.

Ethical Considerations

Responsible AI systems also perform checking from an ethical and fairness perspective, ensuring their decisions are unbiased, transparent, and aligned with human values. Addressing ethical concerns is an integral component of AI system development. This involves mitigating bias, ensuring fairness, and maintaining transparency within the system to promote equitable outcomes and avoid discriminatory behavior.

Additionally, respecting user privacy is of paramount importance, and measures must be implemented to safeguard sensitive data and uphold ethical principles in AI applications. A thorough approach to addressing these ethical considerations helps build trust in AI systems and fosters responsible and ethical AI development.

Step 5: Deployment and Monitoring

Deployment

Upon achieving a satisfactory level of performance, the next step is to deploy the AI model in a real-world environment, allowing it to provide practical solutions. Deployment can take several forms, including integrating the model into existing systems, developing (APIs) for seamless interaction, or creating user interfaces for more accessible usage. This phase is crucial as it transforms the AI model from an experimental concept into a valuable tool that can effectively address real-world challenges and provide practical benefits.

Monitoring and Maintenance

Ongoing monitoring of the AI system's performance and behavior is vital to ensure its reliability and effectiveness in real-world situations. This process involves identifying and addressing issues that may arise, tracking changes in data patterns, and verifying that the system operates as intended. Moreover, the model should be updated as necessary to adapt to shifting data distributions and evolving user requirements, ensuring its continued relevance and value over time. This iterative approach to monitoring and updating guarantees the AI system's ability to consistently provide accurate and useful solutions.

Feedback Loop

Monitoring data serves as a valuable source of feedback, enabling the collection of insights that can guide the necessary updates to the AI model and the overall system. By actively incorporating user feedback and integrating new data, the AI system can continually evolve, enhancing its accuracy, performance, and effectiveness as it adapts to changing conditions and user requirements. This iterative process of improvement ensures that the system remains a valuable and reliable asset over time, meeting the evolving needs of its users and maintaining high performance standards.

Scaling and Optimization

When the AI system's workload demands expansion to handle large-scale operations, it is imperative to optimize it for scalability. This optimization can encompass various strategies such as parallelization, distributed computing, or leveraging cloud services. These methods empower the system to efficiently process higher workloads, ensuring it can accommodate growing demands without compromising performance and responsiveness.

Documentation

Comprehensive documentation of the AI development process ensures transparency and facilitates a thorough understanding of the system's inner workings. This documentation should cover the entirety of the process, encompassing data sources, data preprocessing steps, model architecture, hyperparameters, and deployment details. Not only does this documentation assist in knowledge transfer and enable collaborators to replicate the process, but it also plays a critical role in maintaining the model's integrity and ensuring its reproducibility for future iterations or troubleshooting.

Training and Support

For AI systems that involve user interaction, offering appropriate training and support is crucial for ensuring that users can utilize the system effectively. This may entail user training sessions or the creation of user-friendly guides and documentation that clarify how to navigate and benefit from the AI system. Such support not only enhances the user experience but also optimizes the system's utilization and the overall success of its deployment.

Long-Term Strategy

A successful AI implementation necessitates a forward-thinking approach that extends beyond the initial deployment. It is essential to consider a long-term strategy that encompasses updates, maintenance, and even the potential replacement of the AI model as technology evolves. This proactive approach helps ensure that the AI system remains current, capable, and aligned with the latest technological advancements, allowing it to continue delivering value and staying relevant within its intended domain.

Additional Insights for Building an AI System

Collaboration

Collaboration is pivotal in developing AI systems, uniting multidisciplinary teams comprising data scientists, software engineers, domain experts, and ethicists. This collaborative effort sparks innovation, ensuring that AI systems are designed to tackle complex real-world challenges effectively. Throughout the AI development process, seamless collaboration between multidisciplinary teams is indispensable. This collective approach, pooling expertise, guarantees the effective deployment and operational success of AI systems, which aligns them with the technical requirements and specific needs and goals of the industry or domain they serve.

Continuous Learning

Continuous learning in AI development extends beyond the initial training, involving ongoing exploration as the AI system interacts with real-world data and users. This iterative process empowers AI systems to adapt and refine their behavior over time, enhancing performance and relevance. By actively integrating new insights, AI systems navigate evolving landscapes, ensuring sustained effectiveness in addressing emerging challenges and user behaviors, contributing to their perpetual improvement and versatility.

Adaptability

As AI practitioners encounter unforeseen challenges, they must embrace a willingness to adapt, adjusting their strategies and models to overcome hurdles and enhance the system's capabilities. This flexibility ensures not only the immediate resolution of challenges but also positions AI practitioners to proactively respond to the ever-evolving landscape of technological advancements. Ultimately, the success of AI endeavors relies on the ability to navigate the intricacies of collaboration, continuous learning, and adaptability throughout the development process.

Other Aspects

Beyond collaboration, other crucial aspects permeate the AI development process. Upholding documentation standards, tracking progress diligently, and ensuring compliance with ethical guidelines are imperative steps. Addressing legal and regulatory requirements related to data privacy and AI system usage is equally paramount to prevent legal complications or privacy breaches. These pillars of documentation, ethics, and collaboration are integral, supporting the continued evolution of AI systems and their seamless integration into various domains and industries.

7

Pre-built AI

Ready-to-Use AI

Overview

Ready-to-use AI encompasses prepackaged components, services, and software with built-in AI capabilities or automated algorithmic decision-making. This availability has significantly boosted the adoption of AI, enabling companies to leverage its benefits at a lower cost and in less time. These pre-built AI components, services, and tools are designed for easy integration into applications or systems, providing specific AI capabilities that can be quickly deployed and utilized.

Benefits

Ready-to-use AI solutions are designed to democratize the utilization of artificial intelligence, offering a convenient avenue for individuals, startups, and businesses to tap into AI capabilities without requiring extensive technical expertise. Packaged as user-friendly software applications, accessible APIs, or scalable cloud services, these solutions eliminate the complexities associated with AI development. Users can seamlessly integrate AI into their workflows, leveraging the power of machine learning and automation to enhance processes and decision-making.

The diverse formats of ready-to-use AI, including pre-built software packages for local installations, APIs for easy integration, and cloud-based services for online accessibility, ensure flexibility and cater to a wide range of user preferences. This accessibility accelerates the adoption of AI, allowing users to harness its benefits efficiently and effectively. Whether through local installations for specific use cases, integrated APIs for customized applications, or cloud-based services for scalable resources, the broad spectrum of ready-to-use AI solutions paves the way for a more inclusive and versatile AI landscape.

Ready-to-Use AI Essentials

Ready-to-use AI involves providing pre-built and preconfigured artificial intelligence solutions or services designed to address specific tasks, functions, or industries. It encompasses several key elements, which are described in the following sections.

Prepackaged Solutions

Ready-to-use AI solutions are designed to simplify the integration of artificial intelligence into various applications and workflows. These solutions are typically available in three main formats: pre-built software packages, APIs, and cloud-based services.

Pre-built software packages are user-friendly applications that can be installed on local systems or servers, allowing users to leverage AI capabilities without extensive development efforts. APIs enable developers to seamlessly incorporate AI functionality into their own software. Cloud-based services offer online access to AI resources, ensuring scalability and reducing the need for on-premises infrastructure. These diverse formats cater to a wide range of user preferences and requirements, making AI adoption more accessible and versatile.

Minimal Customization

Users can easily implement these AI solutions with minimal customization or coding expertise, as they often come with user-friendly interfaces and comprehensive documentation. This accessibility means that individuals and organizations with limited coding or AI expertise can harness the power of these technologies without significant development efforts. Ready-to-use AI solutions have democratized access to AI capabilities, enabling a broader range of users to leverage these tools for various applications and industries.

Accessibility

Ready-to-use available AI solutions are frequently accessible through cloud platforms, dedicated marketplaces, or software-as-a-service (SaaS) providers. This accessibility simplifies the procurement process for businesses and individuals, as they can select and deploy AI tools directly from these platforms. Moreover, many providers offer various pricing models, including pay-as-you-go, subscription, and free tiers, increasing flexibility to accommodate diverse user needs and budgets.

Ease of Implementation

Ready-to-use AI solutions are designed to be seamlessly integrated into users' existing systems and applications, streamlining the deployment process. This integration is facilitated through well-documented APIs and Software Development Kits (SDKs), ensuring that developers can quickly leverage AI capabilities without the need for extensive reconfiguration or custom development. As a result, organizations can rapidly harness the benefits of AI without the complexity and time associated with building AI solutions from scratch. This accessibility and ease of integration has democratized the adoption of AI across various industries and use cases.

Use Cases

Ready-to-use AI solutions are designed to cater to specific use cases, ensuring that they excel in their intended applications. For example, natural language processing solutions are optimized for tasks like sentiment analysis, language translation, and text summarization. Image recognition solutions are focused on tasks such as object detection, facial recognition, and scene analysis. Predictive analytics tools are geared toward helping businesses make data-driven decisions and forecasts, while chatbot platforms simplify the development of interactive virtual assistants for customer support or sales. This specialization ensures that users can leverage AI technologies that align precisely with their needs, leading to more efficient and effective solutions.

Industry Focus

Many ready-to-use AI solutions are designed with specific industries in mind, such as healthcare, finance, or e-commerce, and they cater to industry-specific challenges and requirements. This specialization often includes compliance with industry standards and regulations, ensuring that the AI solutions are suitable for their intended use cases. By targeting industry-specific needs, these solutions offer tailored functionalities and support, making them invaluable tools for businesses in those sectors.

User-Friendly Interfaces

Ready-to-use AI solutions often feature user-friendly interfaces and intuitive dashboards, enabling users to configure and manage AI settings without requiring extensive technical expertise. This user-centric design empowers a wider range of individuals, including those without deep technical knowledge, to harness the power of AI in their respective domains. Whether it is fine-tuning chatbot behavior, adjusting recommendation algorithms, or analyzing sentiment in text data, these interfaces make AI accessible to a broader audience.

Overall, ready-to-use AI simplifies the adoption of artificial intelligence by offering pre-built tools and services that cater to specific needs, promoting accessibility and usability across a wide range of applications and industries.

Features of Ready-to-Use AI Solutions

Pretrained Models

Ready-to-use AI solutions typically comprise pretrained machine learning models that have undergone extensive training on substantial datasets, equipping them with the capacity to excel in specific tasks. These models arrive pre-optimized for their intended functions, allowing users to employ them directly for inference tasks, saving valuable time and resources. Moreover, these models can be further fine-tuned to enhance their accuracy and efficiency, tailoring them to meet specific application requirements and ensure optimal performance.

APIs and Libraries

APIs enable developers to interact with AI services and capabilities. They expose functionalities like NLP, image recognition, and sentiment analysis. AI vendors provide APIs and libraries that developers can integrate directly into their applications. Libraries and SDKs provide them with tools and resources to easily integrate AI capabilities into their applications.

Cloud Services

A considerable number of ready-to-use AI solutions are now accessible as cloud-based services, granting users the convenience of online access without the necessity for on-premises infrastructure. This cloud-centric approach simplifies deployment and offers scalability, enabling businesses to adjust resources as needed to accommodate fluctuating demands. Also, cloud-based ready-to-use AI solutions often come equipped with robust security measures and compliance options. These features help safeguard sensitive data and ensure adherence to regulatory standards, a critical consideration in industries where data privacy and regulatory compliance are paramount, such as healthcare and finance.

Pre-built Platforms

Certain platforms provide comprehensive end-to-end solutions tailored to specific applications. A notable example is chatbot platforms that equip users with tools to develop and deploy chatbots without the need to start from the ground up. These platforms often offer pre-built templates, natural language processing capabilities, and intuitive interfaces, simplifying the chatbot creation process and enabling businesses to enhance customer support, engagement, and more without extensive coding efforts.

Minimal Configuration

Ready-to-use AI solutions prioritize user-friendliness, minimizing the complexity of initial setup and configuration. Users can often commence their AI journey promptly by leveraging comprehensive documentation and tutorials provided by the vendor. This streamlined onboarding process ensures that even those with limited technical expertise can harness the power of AI effectively.

Customization Options

Ready-to-use AI solutions strike a balance between convenience and customization. While they come with pre-built capabilities to expedite deployment, they often grant users the flexibility to customize these capabilities to suit their specific use cases. This adaptability empowers organizations to fine-tune AI models, interfaces, and functionalities, ensuring that the AI aligns perfectly with their unique needs and objectives.

Graphical User Interfaces

Graphical User Interfaces (GUIs) in AI simplify the adoption of artificial intelligence for individuals and small businesses lacking coding expertise. These intuitive tools provide a user-friendly environment where users can configure and deploy AI solutions without delving into complex programming tasks. By bridging the technical gap, GUI-based interfaces democratize AI access, enabling a broader audience to leverage the power of AI in various applications.

Scalability

Cloud-based ready-to-use AI services offer scalability advantages, permitting users to easily adjust resources based on fluctuating workloads. This flexibility is particularly valuable for applications with unpredictable demands, as it ensures optimal performance without the need for substantial up-front investments in on-premises infrastructure. As workloads grow or decrease, users can seamlessly scale their AI services up or down, optimizing cost-efficiency and resource allocation.

Domain-Specific Solutions

Many ready-to-use AI solutions are finely tuned to cater to specific industries or specialized use cases, recognizing that different sectors have unique requirements and challenges. Healthcare, for instance, benefits from AI solutions that address medical image analysis, patient diagnosis, and drug discovery. Similarly, the finance sector leverages AI to enhance fraud detection, algorithmic trading, and risk assessment, while customer support relies on chatbots and sentiment analysis tools to improve user experience. These industry-specific AI solutions streamline processes and deliver targeted bene-fits to organizations across various domains.

Data Management Tools

Data management and preprocessing are foundational steps in AI develop-ment, and many AI-ready platforms recognize this importance by offering built-in tools and capabilities. These tools help users handle data efficiently, allowing for cleaning, transformation, and feature extraction tasks. By inte-grating data management features into these platforms, users can streamline their AI projects, ensuring that the data they work with is well prepared for training and analysis.

Pricing Models

AI providers typically offer a range of pricing models for their ready-to-use AI solutions to accommodate diverse user needs. These models often include flexible options like pay-as-you-go, subscription-based plans, and even free tiers for limited usage. This variety allows users to choose the pricing structure that aligns with their specific requirements, whether they need occasional access to AI capabilities or continuous, high volume usage.

Cost-Efficiency

Utilizing ready-to-use AI solutions can be a cost-effective alternative to devel-oping AI applications from the ground up. This approach significantly reduces expenses related to extensive development time, hardware procurement, and the recruitment of AI experts. By leveraging these pre-built tools and services, businesses and individuals can access advanced AI capabilities without incur-ring the substantial up-front and ongoing costs associated with custom AI development.

Democratization of AI Access

The availability of ready-to-use AI democratizes access to advanced artificial intelligence capabilities, breaking down barriers for individuals and organiza-tions without specialized expertise in AI development. This accessibility empowers a broader range of users to harness AI's potential and integrate it into their applications, services, and workflows. By offering user-friendly interfaces, pre-built models, and comprehensive documentation, ready-to-use AI solutions ensure that users from various backgrounds can benefit from the transformative power of artificial intelligence.

Documentation and Tutorials

Ready-to-use AI solutions typically come with comprehensive documentation, tutorials, and guides to facilitate a smooth onboarding experience for users. These resources are invaluable for helping users understand the functionalities and capabilities of the AI tools, ensuring they can quickly and effectively integrate them into their projects. Whether developers are implementing pretrained machine learning models, setting up chatbots, or utilizing natural language processing APIs, well-documented guides play a crucial role in reducing the learning curve and maximizing the benefits of ready-to-use AI.

Versatility of Ready-to-Use AI Solutions

The following sections describe a diverse tapestry of well-known examples that showcase pre-built AI solutions' capabilities, versatility, and convenience.

Language Translation API

Language translation APIs, such as Google Translate, offer developers a convenient way to incorporate robust language translation capabilities into their applications. By utilizing these APIs, developers can bypass the complex and time-consuming process of building their translation models from scratch. Instead, they can seamlessly integrate the translation functionality into their software, expanding the global reach and accessibility of their applications.

Chatbot Platforms

Chatbot platforms provide predesigned chatbots that users can adapt to suit specific purposes such as customer support, sales, or other interactions. This pre-built functionality accelerates the deployment of chatbots, eliminating the need to create them from scratch. Users can fine-tune these chatbots to align with their business objectives and customer needs, enhancing customer engagement and support capabilities.

Virtual Assistant SDKs

SDKs provide developers with the tools and resources needed to create virtual assistants or chatbots capable of understanding and responding to user queries. These kits include pre-built components, libraries, and APIs that simplify the development process, reducing the need for developers to create complex natural language processing and speech recognition systems from scratch. By using SDKs, developers can focus on tailoring virtual assistants to their specific applications and user requirements.

Image Recognition Services

Cloud services such as Amazon Rekognition and Microsoft Azure Computer Vision offer robust image recognition and analysis capabilities suitable for a wide range of applications involving images and videos. These services leverage advanced machine learning models to identify objects, detect faces, extract text, and perform other image-related tasks with high accuracy. Developers can easily integrate these cloud-based services into their applications, eliminating the need to build and train custom image recognition models, accelerating development, and reducing resource requirements.

Image Analysis Services

Ready-to-use image analysis services are designed to analyze images and extract valuable information from them automatically. These services can identify objects, recognize faces, and extract textual content from images, streamlining tasks that involve visual data. Whether it is for content moderation, image tagging, or automating data extraction from images, these ready-to-use solutions offer accuracy and efficiency without the need for extensive custom development.

Speech Recognition APIs

Speech recognition APIs provide developers with the tools to incorporate speech-to-text and voice recognition features into their applications seamlessly. These APIs eliminate the need for developers to create their speech recognition models, saving time and resources. With readily accessible APIs, businesses can enhance their applications with voice-enabled interactions, transcription services, and voice command recognition without the complexities of building speech models from scratch.

Voice Assistants

Platforms such as Amazon Alexa and Google Assistant provide accessible tools and SDKs that empower developers to design and deploy voice-driven applications without delving into the intricacies of creating speech recognition and natural language processing models themselves. These platforms offer pre-built capabilities for voice interaction, allowing developers to focus on crafting engaging and innovative voice-driven experiences for users. By leveraging these tools, businesses and individuals can rapidly bring voice-enabled applications and devices to market, harnessing the potential of voice technology without the need for extensive expertise in AI and speech recognition.

Predictive Analytics Tools

Predictive analytics tools are designed to empower businesses by simplifying the complex process of harnessing data for informed decision-making. They offer user-friendly interfaces and functionalities that enable organizations to utilize data-driven insights without the need for in-depth knowledge of intricate machine learning algorithms. These tools streamline the predictive analytics process by providing pre-built predictive models and automated workflows, making them accessible to a wider range of users and industries, ultimately enhancing data-driven decision-making capabilities.

Text Analysis Services

APIs for text analysis offer developers powerful tools to extract valuable insights from textual data. These APIs provide capabilities such as sentiment analysis to determine emotional tone, named entity recognition to identify important entities like names and locations, and other text-processing tasks like language detection or summarization. By integrating these APIs into their applications, developers can enhance user experiences and gain valuable insights from textual content, making tasks like content curation or social media monitoring easier.

Recommendation Engines

Pre-built recommendation engines are valuable tools for businesses looking to enhance user experiences and boost sales. These engines analyze user behavior, such as browsing history or purchase patterns, to generate personalized product or content recommendations. By offering tailored suggestions, companies can increase customer engagement, encourage repeat visits, and ultimately drive higher conversions, making recommendation engines a key component of e-commerce, content delivery, and other industries reliant on user engagement.

Automated Machine Learning

Automated Machine Learning (AutoML) platforms provide businesses and data scientists with a user-friendly and efficient way to develop machine learning models. They streamline the traditionally complex tasks involved in model training by automating processes like feature selection, hyperparameter tuning, and model evaluation. This automation not only accelerates the development cycle but also makes machine learning accessible to individuals and organizations with varying levels of expertise in data science and AI.

Democratizing AI with Ready-to-Use Solutions

Impact

The impact of ready-to-use AI extends across a multitude of industries and applications. From automating routine tasks to enhancing customer experiences, from improving healthcare diagnostics to optimizing supply chain operations, these solutions empower businesses to innovate, adapt, and thrive in the data-driven era. Moreover, the democratization of AI fosters creativity and experimentation as developers and entrepreneurs explore novel use cases and disruptive innovations that leverage the power of AI. In conclusion, ready-to-use AI is not merely a technological advancement; it is a paradigm shift that promises to reshape industries, empower individuals, and democratize the transformative capabilities of artificial intelligence.

Integration

Ready-to-use AI is a comprehensive and inclusive ecosystem of solutions designed to break down the barriers to entry for artificial intelligence. It transcends the traditional complexities associated with AI development, opening the doors to individuals, businesses, and developers regardless of their level of expertise in the AI domain. This all-encompassing approach comprises prebuilt software packages, APIs, cloud-based services, and specialized platforms tailored to specific industries and use cases. These solutions are engineered to seamlessly integrate into existing systems or applications with minimal customization, streamlining the deployment process significantly.

User-Centric Design

A distinguishing feature of ready-to-use AI offerings is their user-centric design. They often come equipped with user-friendly interfaces, intuitive dashboards, and comprehensive documentation, ensuring that users, regardless of their technical background, can swiftly navigate and harness the full potential of AI. This accessibility is a game changer, making AI capabilities approachable and actionable for a wider audience.

8

Measuring AI Performance

Assessing the Human Likeness of AI

Measuring how closely AI behaves like a human involves evaluating its performance across various dimensions of human-like intelligence. Some ways in which the human likeness of AI behavior can be assessed are described in the following sections.

Task Performance

When assessing the performance of an AI system, it is essential to conduct an evaluation focused on its proficiency in executing tasks demanding human-like skills. An example is natural language processing, where AI's capability to comprehend and generate text responses in a manner reminiscent of human communication becomes a key metric. This evaluation should encompass comprehensive testing and analysis to gauge AI's effectiveness in mimicking human-like skills and its aptitude for generating contextually relevant and linguistically coherent responses.

Accuracy and Precision

The accuracy and precision of an AI system's outputs serve as crucial metrics in evaluating its human likeness. When the system's answers or actions closely mirror what a human would produce in similar situations, it reflects a higher degree of human likeness. This evaluation entails rigorous testing to determine the AI's capacity for delivering precise and reliable outcomes consistent with human expectations.

Contextual Understanding

Assessing the AI's capacity to understand and respond appropriately in diverse contexts is vital to measuring its human likeness. Similar to how humans adapt their responses based on the situation, AI's ability to recognize and adapt to different contexts signifies a higher level of human likeness. This evaluation involves testing the AI system in various scenarios to ensure it can provide contextually relevant and nuanced responses.

Adaptability

Evaluating AI's capability to adapt to new or unexpected situations is crucial for assessing its human likeness. Human-like behavior often entails navigating unfamiliar scenarios, learning from experiences, and making informed decisions in real-time. This assessment involves exposing the AI to unscripted and unpredictable situations to gauge its capacity for adaptation and decision-making in a manner akin to human responses.

Common Sense Reasoning

It is important to evaluate its ability to exhibit common sense reasoning and logical inferences akin to human thinking in everyday scenarios to assess the human likeness of AI. This involves testing AI's capacity to draw logical conclusions from given information, make sound judgments, and apply common-sense knowledge to solve problems. Evaluating AI's performance in such tasks provides insights into its cognitive abilities and how closely it resembles human-like reasoning.

Naturalness

When evaluating the human likeness of AI, it is crucial to consider the naturalness of its interactions. Human-like behavior entails using natural language fluently, employing appropriate gestures, and conveying tone effectively in communication. Assessing these aspects helps determine AI's ability to engage with users in a manner that closely resembles human interaction, contributing to its overall human likeness.

Emotional Understanding

Evaluating AI's human likeness involves assessing its ability to recognize and appropriately respond to emotions, including empathy, sympathy, and understanding user sentiment. AI's ability to perceive and react to emotional cues is crucial to human-like behavior. The closer an AI system comes to replicating human emotional intelligence, the higher its human likeness can be considered.

Creativity and Imagination

Assessing the human likeness of AI includes evaluating its ability to generate creative content, ideas, or solutions that showcase human-like imagination. This involves determining whether AI can go beyond preprogrammed responses and generate novel, innovative, and imaginative outputs similar to human creativity, thereby contributing to a more comprehensive understanding of its capacity to emulate human cognitive processes in the realm of creative thinking and problem-solving.

Ethical and Moral Considerations

Evaluating the human likeness of AI also involves gauging whether the AI system adheres to ethical principles and moral values in its decision-making, aligning with human ethical behavior. This assessment includes considering whether the AI system makes choices consistent with established ethical guidelines and societal norms, just as humans do in ethical dilemmas. The capacity for AI to exhibit ethical reasoning and decision-making contributes to its human likeness, particularly in contexts where ethical considerations are crucial.

Learning and Adaptation

It is important to measure how well AI learns from new information and experiences, evolving its behavior over time in a manner akin to human learning, to assess the human likeness of AI. This evaluation examines AI's ability to adapt and improve its performance based on new data and feedback, mirroring how humans acquire knowledge and refine their skills through continuous learning. The capacity for AI to exhibit learning mechanisms similar to humans underscores its potential human-likeness and adaptability in dynamic environments.

Bias and Fairness

For evaluating the human likeness of AI, it is crucial to analyze AI's decision-making for biases and fairness. This assessment involves scrutinizing AI's responses and actions to ensure they align with human values and do not exhibit discriminatory behavior. By examining AI's capacity to make impartial and ethically sound decisions, we can better understand its alignment with human-like ethical judgment and fairness principles.

Error Handling

Assessing AI's response to errors or unexpected inputs is essential for gauging its human likeness. A truly human-like AI should be capable of acknowledging errors gracefully and providing appropriate responses when faced with unexpected or incorrect inputs. This evaluation helps determine AI's ability to handle situations in a manner reminiscent of human adaptability and understanding.

Long-Term Planning

Evaluating AI's ability to engage in long-term planning and goal setting is crucial for measuring its human likeness. A human-like AI should demonstrate the capacity to consider future consequences and implications when making decisions and setting goals. This assessment helps determine if AI exhibits forward-thinking behavior similar to human beings, showcasing its strategic thinking and foresight capacity.

Conversational Depth

Assessing the depth and complexity of AI's conversations is essential to determine its human likeness. Human conversations often involve nuanced and multifaceted discussions that require understanding context, providing detailed responses, and addressing various aspects of a topic. By evaluating AI's ability to engage in such intricate conversations, we can better gauge its capacity to emulate human communication.

Social Interactions

Evaluating how well AI behaves in social interactions is vital for assessing its human likeness. Human social interactions involve a range of behaviors, such as politeness, empathy, and adapting to various social contexts. By examining AI's ability to exhibit these qualities, we can gain insights into its capacity to emulate human-like social interactions.

It is important to note that achieving human-level behavior across all these dimensions is extremely challenging due to the complexity of human intelligence. AI's primary goal is to be functional, efficient, and valuable rather than exactly replicating human behavior. Measuring human likeness can help developers and researchers understand where AI excels and where improvements are needed, but it is essential to set realistic expectations based on the specific capabilities and limitations of AI technologies.

Methods for Measuring AI

Measurement Methods

The following measurement methods are used to measure how closely AI is behaving like a human:

- Turing Test
- Cognitive Modeling Approach
- Laws of Thought Approach
- Rational Agent Approach

Turing Test

Proposed by Alan Turing in 1950, the Turing Test involves a human judge engaging in natural language conversations with both a human and a machine. The basis for this test is that the AI entity should be able to conduct a conversation with a human. Ideally, the human judge should not be able to conclude that he is talking to an AI entity. For this, AI needs to possess the following qualities:

- Natural Language Processing to communicate successfully
- Knowledge Representation, which acts as its memory

- Automated Reasoning, which uses stored information to answer questions and draw new conclusions
- Machine Learning to detect patterns and adapt to new circumstances

If the judge cannot reliably distinguish between the two based on their responses, the machine is said to have passed the test and demonstrated human-like intelligence. The Turing Test emphasizes the ability of AI to mimic human conversation and behavior convincingly. However, passing the test does not necessarily mean AI understands or thinks like a human; it just exhibits human-like responses.

In a simplified explanation, the Turing Test is like a game or a test for machines, especially computers, to see if they can think and talk like humans. Imagine you are chatting with someone over a computer, but you cannot see them. You can only type messages back and forth. In this test, there are three participants: you, another person, and a computer. The goal is for you to fig-ure out which one of the other two is the computer just by chatting with them.

If the computer can make you think it's the person, then it "passes" the Turing Test. This means it can communicate so well that you cannot tell it is a machine. It is like a test of how smart a computer is when it comes to having conversations. The Turing Test helps us understand if a computer can talk and think like a human. It is a way to measure how advanced artificial intelligence has become in mimicking human communication.

Cognitive Modeling Approach

Cognitive models are a possible bridge between behavior and the brain. A cognitive model is an approximation of one or more cognitive processes in humans for the purposes of comprehension and prediction. They embody psychological principles and are often evaluated by their ability to account for behavioral data.

The Cognitive Modeling Approach involves creating computational mod-els that simulate human cognitive processes, aiming to replicate the information-processing mechanisms and behaviors observed in human cogni-tion, including memory, perception, learning, and decision-making. This approach focuses on understanding and replicating the underlying cognitive mechanisms of human thinking, seeking to build models that not only mimic human behavior but also provide insights into how humans process informa-tion. The ultimate goal is to distill the essence of the human mind and build AI models based on human cognition.

To achieve this, there are three main approaches:

- Introspection: Observing our thoughts and building a model based on it
- Psychological experiments: Conducting experiments on humans and observing their behavior
- Brain imaging: Using MRI to observe how the brain functions in different scenarios and replicating that through code

In a simplified explanation, the AI cognitive modeling approach is like trying to make a computer think and learn the way humans do. For example, when we want to teach a computer to do something, we often break it down into smaller steps or rules. However, the cognitive modeling approach is different. It tries to make the computer think more like a person. Therefore, instead of just giving the computer a set of instructions, we try to make it understand things, remember information, and learn from its experiences, just like we do.

If you want to teach a computer to recognize cats, instead of telling it, "Look for pointy ears, fur, and a tail," you might try to make it learn by showing it numerous pictures of cats and other animals so that it can figure out the differences on its own.

In summary, the cognitive modeling approach is all about developing AI systems that can think, learn, and problem-solve more like humans, using processes similar to how our brains work. It is a way to create smarter and more adaptable AI.

Laws of Thought Approach

The Laws of Thought Approach, often associated with classical symbolic AI, involves formalizing human reasoning using symbolic logic and mathematical rules. It seeks to develop systems that adhere to well-defined logical principles, such as deduction and inference.

This approach is like using a set of strict rules to make a computer think logically and make decisions. It is rooted in formal logic and attempts to capture human reasoning using explicit rules and representations. It emphasizes the systematic manipulation of symbols to arrive at valid conclusions. These rules can be thought of as simple principles that a computer follows to come up with answers.

There are three main laws of thought:

1. The Law of Identity:
   It says that something is always itself. For example, if you have an apple, it is always going to be that same apple, no matter what.
2. The Law of Non-contradiction:
   It means that two opposite things cannot be true at the same time. Like, you cannot have a sunny day and a rainy day at the exact same moment.
3. The Law of the Excluded Middle:
It says that something is either true or false; there is no middle ground. For instance, if we are talking about whether it is raining outside, it is either true (it is raining) or false (it is not raining); there is no middle ground.

With the laws of thought, computers use these rules to think logically and make decisions without any ambiguity or contradictions. It is a very precise way of approaching problems, making AI systems follow strict rules to make choices and draw conclusions.

It should be realized that a large list of logical statements governs the operation of our mind, which can be codified and applied to AI algorithms. However, there is an issue with this approach. Solving a problem in principle strictly according to the laws of thought and solving them in practice can be quite different, requiring contextual nuances to apply. Also, humans take some actions without being 100% certain of an outcome, which an algorithm might not be able to replicate if there are too many parameters.

Rational Agent Approach

The Rational Agent Approach views AI as rational agents that perceive their environment and take actions to maximize their expected utility. It focuses on decision-making based on reasoning and optimization, considering the agent's goals and available information.

This approach is grounded in the concept of rational decision-making. It emphasizes AI's ability to make optimal choices given its knowledge and objectives. It encompasses a wide range of AI paradigms, including classical AI, machine learning, and more.

The rational agent approach in AI is like teaching a computer to be really smart and make good decisions. Imagine the computer as an "agent" that wants to do the best it can in a certain situation. Just like you want to make the best choices in life, this computer agent wants to make the best choices in its world. For this, it gathers information about its environment, thinks about what actions it can take, and then picks the action that will lead to the best outcome. It is a bit like a robot trying to figure out the best way to complete a task.

This approach is called "rational" because the computer tries to be as logical and clever as possible. It doesn't always have to follow strict rules like the "Laws of Thought" discussed earlier. Instead, it aims to make the most sensible decisions based on the information it has. Therefore, the rational agent approach is all about creating AI systems that act like smart, thoughtful beings, making choices that lead to the best results in different situations.

A rational agent aims to maximize its expected utility based on its beliefs, preferences, and available information, seeking the best possible outcome given its current circumstances. While classical logic principles, known as the "Laws of Thought," play a role in rational decision-making, it is important to note that rational agents navigate the complexities of decision-making, especially when faced with multiple potential outcomes and necessary compromises. Rather than seeking a "logical right thing to do" in an absolute sense, they make decisions based on logical inferences from their beliefs and evidence. This approach makes rational agents dynamic and adaptable, allowing them to update their strategies in response to changing circumstances and uncertainties.

9

Comparing Measurement Methods

Approaches

A number of approaches exist for comparing measuring methods, which are described in the following sections. Each approach has its strengths and limitations, and the choice of approach often depends on the specific goals, applications, and philosophical viewpoints of AI researchers and practitioners.

Emphasis on Behavior

The Turing Test and Cognitive Modeling Approach are two distinct methods aimed at achieving human-like behavior in artificial intelligence. The Turing Test primarily focuses on evaluating an AI system's conversational ability, challenging it to engage in a conversation indistinguishable from that of a human. In contrast, the Cognitive Modeling Approach goes beyond conversation and aims to simulate the cognitive processes involved in human decision-making and problem-solving.

While the Turing Test evaluates the surface-level behavior of AI, the Cognitive Modeling Approach delves deeper into replicating human-like thought processes and reasoning, ultimately striving for a more comprehensive emulation of human behavior in AI systems.

© The Author(s), under exclusive license to Springer Nature Switzerland AG 2024
A. Khan, *Artificial Intelligence: A Guide for Everyone*,
https://doi.org/10.1007/978-3-031-56713-1_9

Formal Logic Versus Rationality

The Laws of Thought Approach and the Rational Agent Approach are two distinct paradigms within artificial intelligence. The Laws of Thought Approach relies heavily on formal logic and symbolic representation to model and solve problems. It uses logical rules and symbolic manipulation to emulate human thought processes. In contrast, the Rational Agent Approach strongly emphasizes rational decision-making, where AI systems make choices based on defined goals and utility functions. It is more focused on achieving optimal outcomes by considering the consequences of various actions.

While the Laws of Thought Approach is rooted in logic and symbolic reasoning, the Rational Agent Approach prioritizes goal-driven decision-making, showcasing the different paths AI can take in achieving the desired objectives.

Cognitive Versus Rational Perspective

The Cognitive Modeling Approach and the Rational Agent Approach represent distinct perspectives in artificial intelligence. The Cognitive Modeling Approach aims to delve deeply into the intricacies of human cognition, striving to replicate and understand how humans think and process information. In contrast, the Rational Agent Approach is primarily concerned with AI systems making rational decisions based on predefined goals and utility functions, regardless of whether these processes directly mimic human cognition. While the former focuses on emulating human thought processes, the latter prioritizes achieving optimal outcomes through rational decision-making, often considering factors beyond human cognitive patterns.

Range of Approaches

The Rational Agent Approach provides a versatile and encompassing framework with the flexibility to incorporate a wide array of methods and techniques from the field of artificial intelligence. It serves as a foundational concept that can accommodate diverse AI methodologies, ranging from symbolic AI, which relies on formal logic and rule-based systems, to contemporary approaches like neural networks, deep learning, reinforcement learning, and many others. This adaptability makes the Rational Agent Approach a powerful and inclusive paradigm for designing AI systems that can make rational decisions and achieve specific goals across a multitude of domains, reflecting the ever-evolving landscape of AI research and development.

Additional AI Assessment Methods and Metrics

Several other methods and metrics exist for assessing how closely AI behaves like a human. These methods aim to evaluate various aspects of AI behavior and capabilities. Collectively, these methods provide a more comprehensive assessment of AI's behavior, intelligence, and capabilities beyond the Turing Test. Each method focuses on specific aspects of human-like behavior, and their combined use offers a more holistic understanding of AI's strengths and limitations. The notable ones are described in the following sections.

Winograd Schema Challenge

The challenge in question involves a collection of sentences deliberately crafted with ambiguous pronouns, tasking the AI system with the intricate task of discerning the accurate referent. This challenge serves as a litmus test for AI's proficiency in grasping subtle contextual cues and effectively resolving linguistic ambiguities, mirroring the crucial facets of human-like language comprehension. By navigating the intricacies of ambiguous pronouns, AI demonstrates its capacity to interpret and respond to text with a higher level of contextual awareness and precision, a significant benchmark in advancing the capabilities of natural language understanding in artificial intelligence.

CAPTCHA Tests

CAPTCHAs, an acronym for "Completely Automated Public Turing test to tell Computers and Humans Apart," tests to tell computers and humans apart. It serves the dual purpose of distinguishing between human users and automated bots, while also serving as an evaluation tool for artificial intelligence capabilities. These challenges, with their focus on visual perception and pattern recognition, extend beyond merely thwarting bots. They stand as a test bed to gauge AI's competence in tasks that necessitate discerning intricate visual elements and recognizing patterns, demonstrating the evolving sophistication of AI systems in mimicking human-like cognitive functions.

Image Recognition and Classification Benchmarks

AI models undergo rigorous evaluation to assess their proficiency in accurately recognizing and categorizing objects depicted in images. Benchmark assessments like ImageNet serve as critical yardsticks, measuring how much AI's image recognition capabilities converge with human-level performance. These evaluations play a pivotal role in gauging the advancements made in computer vision, helping researchers and developers understand the strengths and limitations of AI models in tasks that parallel human visual comprehension.

Commonsense Reasoning Challenges

The "Commonsense Reasoning Challenge" and similar assessments pose questions to AI systems that demand commonsense reasoning for accurate responses. These challenges serve as litmus tests to evaluate AI's capacity to emulate human-like thinking and draw logical inferences grounded in everyday knowledge. By probing AI's ability to navigate the intricacies of common sense, these challenges shed light on the system's progress in comprehending and reasoning about the world in a manner akin to human cognition.

Reading Comprehension Tasks

Reading comprehension assessments, such as the Stanford Question Answering Dataset (SQuAD), tasks AI systems to read a given passage and provide accurate answers to questions about that text. These evaluations serve as robust measures of AI's reading comprehension capabilities, as they demand not only the understanding of textual content but also the ability to extract and synthesize information effectively. By participating in tasks like SQuAD, AI models demonstrate their capacity to interpret and derive meaningful insights from textual data, mirroring the cognitive skills used by humans in reading and comprehension.

Emotion Recognition Tests

Emotion recognition datasets are instrumental in assessing AI systems' competence in discerning human emotions conveyed through text or voice. These evaluations provide a means to measure AI's emotional intelligence, requiring it to accurately identify and categorize the emotional tone expressed by individuals. By participating in such tasks, AI models demonstrate their capacity to process linguistic or auditory information and interpret the subtle nuances and sentiments that are integral to human communication, bridging the gap between artificial and human emotional understanding.

Ethical Decision-Making Scenarios

Evaluating AI's ethical behavior involves subjecting it to scenarios that simulate ethical dilemmas. The AI's responses and decision-making processes are meticulously scrutinized in these scenarios to gauge how closely they align with the ethical considerations and moral reasoning that humans typically apply in similar situations. Such assessments are crucial in ensuring that AI systems adhere to ethical principles and make decisions that are in harmony with human values, thereby fostering trust and responsible use of artificial intelligence in various applications and industries.

Conversational Depth and Cohesion

AI's ability to maintain coherent and contextually relevant conversations is evaluated through chatbot-based assessments. These assessments focus on measuring the quality and depth of dialogues initiated by AI systems. They examine how well AI can engage in meaningful and context-aware conversations, understand user queries, and provide appropriate responses. These evaluations are vital for determining the conversational capabilities of AI chatbots and ensuring they can interact effectively with users across various domains, from customer support to virtual assistants, ultimately enhancing user satisfaction and engagement.

Collaborative Problem-Solving

Evaluating how effectively AI collaborates with humans in problem-solving scenarios encompasses various aspects. It involves assessing AI's ability to not only contribute relevant insights but also adapt to different human communication styles, ensuring effective interaction and cooperation. Additionally, measuring AI's collaborative performance includes evaluating its capacity to work harmoniously toward a common goal with human counterparts, fostering synergy in tackling complex problems. These assessments play a crucial role in determining AI's utility in collaborative contexts, such as team-based decision-making and group problem-solving scenarios, where seamless interaction between AI and humans is paramount for achieving optimal outcomes.

Transfer Learning Performance

The evaluation of AI's capacity to generalize its knowledge to novel, unseen scenarios is a fundamental test of its human-like cognitive abilities. This assessment assesses AI's proficiency in applying learned concepts and principles across a diverse range of contexts, mirroring the way humans adapt their knowledge and skills to tackle new challenges. In this context, generalization underscores AI's ability to transcend the limitations of rigid, context-specific learning and instead exhibit flexible and adaptable intelligence akin to human cognition. This capability is crucial for AI systems to handle real-world situations effectively and contribute to tasks that require the application of acquired knowledge in novel and unforeseen circumstances.

Long-Term Planning and Goal Achievement

The assessment of how AI engages in long-term planning and goal-setting tasks is instrumental in gauging its ability to replicate human-like strategic thinking. This evaluation delves into AI's proficiency in formulating and executing plans that span extended periods, accounting for various potential outcomes, and adapting strategies to achieve predefined objectives. By scrutinizing its performance in these scenarios, insight is gained into AI's aptitude for considering future consequences, optimizing its actions over time, and navigating complex decision-making landscapes with a strategic foresight akin to human cognitive processes. This capability is paramount in various applications, such as autonomous systems, business strategy optimization, and game playing, where forward-thinking and goal-oriented behavior are prerequisites for success.

Bias and Fairness Analysis

The assessment of AI's adherence to human values and fairness revolves around scrutinizing its outputs for biases and discriminatory behavior. This evaluation involves examining how the AI system interacts with data and makes decisions, ensuring that it aligns with ethical principles and does not exhibit partiality or unfairness toward individuals or groups. Detecting and mitigating biases in AI's decision-making processes is crucial, particularly in applications where the impact on human lives and society at large is significant, such as lending, hiring, justice, and healthcare. By thoroughly analyzing its outputs and ensuring they reflect a fair and just approach, it can be ascertained that AI systems are in harmony with the moral and ethical standards upheld by human societies.

10

Simulating Intelligence

Cognitive Skills

Cognitive Skills for AI

Artificial intelligence emphasizes three cognitive skills: learning, reasoning, and self-correction—skills that the human brain possesses to one degree or another. These three cognitive skills are essential for making AI systems smarter and more capable of handling complex tasks. They allow AI to adapt, make better decisions, and continually improve their performance. In the context of AI, these skills are explained in the following sections.

Learning

This refers to the acquisition of information and the rules needed to use that information. Learning in AI is like teaching a computer to get better at a task by providing it with data and experiences. It is a bit like how we learn from our experiences. For example, if you want a computer to recognize cats in pictures, you will show it lots of cat images. Over time, it learns to spot common features of cats and can recognize them in new pictures. This is called machine learning, and it is a fundamental skill for AI.

© The Author(s), under exclusive license to Springer Nature Switzerland AG 2024
A. Khan, *Artificial Intelligence: A Guide for Everyone*,
https://doi.org/10.1007/978-3-031-56713-1_10

Reasoning

This refers to using information rules to reach definite or approximate conclusions. Reasoning is about using information to come up with conclusions or make decisions. It is like solving puzzles or making logical choices. In AI, reasoning involves taking what the computer has learned and using it to make smart decisions. For instance, if a self-driving car has learned the rules of the road and observed its surroundings, it can decide when to stop at a red light or change lanes safely.

Self-correction

This involves the process of continually fine-tuning AI algorithms and ensuring that they offer the most accurate results. Self-correction is a bit like learning from your mistakes. AI systems are not perfect, and they can make errors. Self-correction involves AI recognizing when it is wrong and trying to improve. For instance, if a chatbot gives the wrong answer, it can analyze why it made the mistake and learn not to repeat it. This skill helps AI become more accurate and reliable over time.

Problem-Solving Techniques

Overview

AI researchers have adapted and integrated a wide range of problem-solving techniques to make artificial intelligence smarter and more versatile. These include search and mathematical optimization, formal logic, artificial neural networks, and methods based on statistics, probability, and economics. AI draws upon computer science, psychology, linguistics, philosophy, and many other fields. A simplified overview of how they have done it is described in the following sections.

Rule-Based Systems

During the nascent stages of AI development, researchers predominantly employed rule-based systems as their primary approach. These systems operated on a fixed set of predefined rules that dictated decision-making processes.

For instance, a medical expert system might have rules such as "if a patient exhibits symptoms like fever and a sore throat, it might be indicative of the flu." However, the drawback of these rule-based systems was their inherent limitation to the specific tasks and scenarios they were explicitly programmed for, lacking the ability to adapt or generalize beyond their predefined rulesets.

Machine Learning

AI researchers then turned to machine learning. Instead of relying solely on predefined rules, machine learning algorithms allow AI systems to learn from data. For instance, in image recognition, AI learns to identify objects by analyzing thousands of images with known labels. This approach enables AI to recognize patterns and make predictions based on what it has learned.

Neural Networks

Within machine learning, neural networks have been a game changer. They are inspired by the structure of the human brain and consist of interconnected nodes (artificial neurons) organized in layers. Neural networks are particularly good at tasks like image and speech recognition, language processing, and game playing. Deep learning, which involves deep neural networks with many layers, has achieved remarkable results in various AI applications.

Natural Language Processing

Natural language processing (NLP) techniques are at the core of AI's ability to comprehend and produce human language. Through NLP integration, AI facilitates conversational chatbots, seamless translation services, and insightful sentiment analysis. This technological synergy enhances AI's capacity to engage with humans in a genuine and intuitive manner. Furthermore, NLP empowers AI to interpret textual data, extract insights, and contribute to various language-related tasks, ranging from content summarization to language generation.

Reinforcement Learning

This approach to AI learning can be likened to a system of rewards and punishments, mirroring the way humans and animals learn from their experiences. AI agents continuously learn by taking actions within a defined environment and receiving feedback based on their performance. For instance, consider the training of a self-driving car: it learns to navigate the roads safely by being rewarded for adhering to traffic rules and penalized for accidents or violations. This system of reinforcement learning allows AI to iteratively refine its decision-making processes, striving for actions that maximize rewards while minimizing penalties, ultimately leading to more intelligent and adaptive behavior.

Evolutionary Algorithms

Inspired by biological evolution's mechanism, these algorithms have found applications across diverse domains, serving to optimize solutions and improve efficiency. They have been particularly valuable in fields like aerospace engineering, where they aid in the design of more aerodynamically efficient aircraft wings, as well as in software development, where they assist in creating more streamlined and effective computer programs. These evolutionary algorithms' adaptability and problem-solving capabilities make them powerful tools for tackling complex optimization challenges.

Hybrid Approaches

Researchers often employ a hybrid approach in practical scenarios by integrating multiple AI techniques to harness their respective strengths. Consider the example of a self-driving car: it may employ rule-based systems for safety-critical decisions, ensuring that fundamental traffic regulations are consistently followed. Simultaneously, it can leverage deep learning algorithms to recognize and respond to dynamic elements such as objects and pedestrians, harnessing the power of neural networks to process complex visual data efficiently. This amalgamation of rule-based and deep learning systems exemplifies the synergy achieved when combining various AI methodologies to enhance the performance and robustness of AI applications.

Continuous Improvement

The field of AI is characterized by its dynamic and ever-evolving nature, with researchers consistently fine-tuning established techniques while pioneering novel approaches. Concurrently, there is a strong emphasis on creating AI systems that can elucidate their decision-making processes, promoting transparency and accountability. Ethical considerations are at the forefront of AI development, as researchers strive to imbue AI systems with ethical guidelines to ensure responsible and fair use, and they are working on enhancing AI's adaptability to thrive in diverse and shifting environments.

In summary, AI researchers have adapted and integrated a wide range of problem-solving techniques, from rule-based systems to cutting-edge deep learning, to make AI systems more capable, flexible, and useful in a variety of applications. The field of AI is dynamic, and it is likely to see even more innovations in the future.

Creating or Simulating Intelligence

Sub-problems

The general problem of simulating or creating intelligence has been broken down into several sub-problems to make it more manageable and approachable.

The machine learning sub-problem focuses on teaching computers to learn from data. It involves developing algorithms and techniques that allow AI systems to recognize patterns, make predictions, and improve their performance over time. Machine learning includes supervised learning (where AI learns from labeled data), unsupervised learning (where it learns without labeled data), and reinforcement learning (where it learns through trial and error).

Natural language processing deals with enabling AI to understand, generate, and interact with human language. Sub-problems within NLP include speech recognition (converting spoken language to text), language understanding (comprehending the meaning of text or speech), language generation (creating human-like text or speech), and language translation.

Besides machine learning and natural language processing, there exist other sub-problems for creating or simulating intelligence, which are described in the following sections.

Computer Vision

This sub-problem focuses on allowing AI systems to "see" and interpret visual information from images or videos. It involves tasks such as image recognition (identifying objects in images), object tracking (following objects in videos), and image generation (creating images from descriptions).

Robotics

Robotics encompasses the development of tangible machines capable of engaging with the physical world. Within this field, critical sub-problems demand attention, including motion planning, which involves determining the optimal movement strategy for robots to achieve specific objectives. Sensor integration plays a vital role by enabling robots to process data from various sensors such as cameras and touch sensors, while human-robot interaction seeks to enhance robots' capacity to comprehend and respond to human actions and speech, facilitating seamless collaboration in various domains.

Knowledge Representation

The sub-problem concerning knowledge representation within AI revolves around the effective storage and utilization of information by AI systems. It entails the design and implementation of databases and data structures that facilitate the efficient storage and retrieval of knowledge. Techniques such as ontologies and semantic networks are employed to provide structured and interconnected representations of information, enabling AI systems to make sense of data and draw meaningful insights from it.

Reasoning and Problem-Solving

Reasoning and problem-solving form the bedrock of AI's cognitive capabilities, enabling it to navigate the intricacies of complex tasks and scenarios. To achieve this, AI addresses various sub-problems that empower it to excel in logical thinking and decision-making. Automated reasoning is a critical component, imbuing AI systems with the ability to make logical deductions, thus enhancing their capacity to derive meaningful conclusions from data.

Expert systems represent a specialized branch of AI that equips machines with the proficiency to tackle domain-specific, expert-level challenges, mirroring the expertise of human specialists. Complementing these capabilities, planning mechanisms empower AI to chart out sequences of actions that lead to the attainment of predefined goals, ensuring AI's adaptability and effectiveness in a multitude of applications.

Ethics and Fairness

With the increasing potency of AI, an imperative arises to confront ethical considerations and uphold principles of fairness and responsibility. Within this realm, sub-problems take center stage, encompassing the formulation of comprehensive guidelines and algorithms aimed at fostering ethical AI conduct. These initiatives are designed to not only curtail biases within AI systems but also safeguard individual privacy, thereby ensuring that AI technologies align with societal values and adhere to stringent ethical standards.

Meta-learning and Transfer Learning

The realm of AI extends into the challenge of enabling machines to acquire the skill of learning itself, leading to the sub-problems of meta-learning and transfer learning. Meta-learning is a specialized field that centers on enhancing AI's capacity to swiftly adapt to novel tasks by building upon prior learning experiences. This not only bolsters AI' agility but also enables it to glean valuable insights from past tasks to excel in future endeavors. On the other hand, transfer learning empowers AI to leverage knowledge acquired in one domain to enhance its performance in another, fostering a broader application of AI's capabilities and enabling cross-disciplinary problem-solving. These facets of AI research are pivotal in advancing the adaptability and versatility of AI systems in an ever-evolving landscape of tasks and challenges.

Explainability and Interpretability

In many AI applications, there arises a necessity for AI systems to elucidate their decision-making processes to human users. Within this context, the associated sub-problems revolve around crafting methodologies and techniques that enhance the transparency and comprehensibility of AI-generated decisions and predictions. By addressing these challenges, AI developers seek to establish a bridge of understanding and trust between humans and AI, ultimately promoting responsible and effective integration of AI technologies in diverse domains.

By breaking down the general problem of simulating intelligence into these sub-problems, researchers can tackle specific challenges and make steady progress in advancing the field of AI. Each sub-problem requires its own set of techniques, algorithms, and research efforts, contributing to the overall development of intelligent machines.

Objectives of AI Research

Traditional Goals

Traditionally, AI research has pursued several core goals and objectives, which have evolved over time as the field has developed. General intelligence, the ability to solve an arbitrary problem, is among AI's long-term goals. The general problem of simulating (or creating) intelligence has been broken down into sub-problems. These consist of particular traits or capabilities that researchers expect an intelligent system to display.

The traditional goals of AI research, listed below, are described in detail in subsequent chapters:

- Reasoning
- Knowledge representation
- Planning
- Learning
- Natural language processing
- Perception
- Motion and manipulation
- General intelligence

The following section describes AI research goals from a different angle.

AI Research Goals: Alternative Perspective

Computer Vision

Researchers embarked on a quest to endow machines with the capability to comprehend and interpret visual information derived from images and videos, seeking to bridge the gap between artificial intelligence and human perception. Some areas within computer vision are object recognition, scene understanding, and image generation.

Robotics

The pursuit of creating physically embodied AI systems with the capability to engage with the tangible world has been a fundamental objective in the realm of artificial intelligence. This encompasses tasks such as robot motion control, navigation, perception (sensing the environment), and manipulation (handling objects).

Expert Systems

In the nascent stages of AI research, the primary objective was the development of expert systems, sophisticated computer programs meticulously designed to replicate the decision-making acumen exhibited by human experts in specialized domains. These systems aim to solve complex problems by emulating human expertise.

Machine Creativity

Some researchers have been actively pursuing the goal of determining whether AI systems can display creative thinking, including tasks like generating art, music, or literature, as this capability could potentially lead to groundbreaking advancements in various creative domains.

Autonomous Agents

Developing AI agents capable of operating autonomously in complex environments, making decisions, and adapting to changing conditions without human intervention has been a major objective, especially in fields like autonomous vehicles and drones.

Ethical AI

Ensuring that AI systems behave ethically, fairly, and responsibly has become an increasingly important goal as AI technologies have become more pervasive. Researchers are working to develop AI that respects human values and societal norms.

Evolution of AI Research Goals

Traditional Objectives: Historical Perspective

In the early stages of AI research, the objectives were tightly focused on achieving specific, often narrow, tasks. This historical perspective reveals a landscape where the nascent field grappled with limited technological capabilities, and researchers aimed to conquer individual challenges within these constraints. The emphasis was on laying the foundation for what would become a transformative journey in AI research.

AI researchers initially sought to replicate human intelligence in narrowly defined domains as technological innovation progressed. The objectives revolved around developing systems that could excel in tasks such as chess playing, theorem proving, and language translation. While groundbreaking, these early pursuits underscore the historical context of AI research and set the stage for broader evolution.

Modern Landscape: Versatility and Comprehensiveness

In the contemporary era, AI research has undergone a profound transformation, marked by a departure from isolated goals towards the integration of versatile and comprehensive functionalities. Researchers today aspire to create AI systems that go beyond narrow, task-specific applications, embracing a holistic approach to address the complexities of the modern world effectively.

This shift in focus emphasizes the development of AI systems with broad-ranging capabilities. From natural language processing and computer vision to complex decision-making, modern AI research seeks to create systems that can adapt and excel across diverse tasks. The multifaceted nature of the modern AI landscape is characterized by a synthesis of various objectives and strategies, aligning with the ever-evolving needs of society and technology.

The adaptability inherent in contemporary AI research reflects a dynamic response to the challenges posed by the intricate demands of our present era. As AI continues to progress, researchers are not only pushing the boundaries of technological innovation but also shaping a field that is increasingly intertwined with the complex and evolving intersections of society and technology.

11

Traditional Goals of AI Research

Reasoning and Problem-Solving

Overview

Reasoning and problem-solving are fundamental components of artificial intelligence (AI), enabling machines to analyze intricate situations, identify challenges, and generate effective solutions to achieve specific objectives. These processes involve logical thinking, deduction, and inference, allowing AI systems to make informed decisions based on available information.

AI utilizes a diverse range of techniques, including rule-based reasoning, search algorithms, and optimization methods, to navigate complex problem spaces. The ultimate goal is to enhance decision-making, problem resolution, and adaptability, making AI systems invaluable across a wide range of domains. The application of reasoning and problem-solving extends across various domains, from healthcare and finance to robotics and autonomous systems, enhancing decision-making, problem resolution, and overall adaptability in AI systems.

Process

The reasoning and problem-solving process in AI constitutes a foundational element in the realm of artificial intelligence, enabling machines to navigate intricate scenarios with competence. It involves more than just processing data; it encompasses the ability of machines to meticulously analyze

A. Khan, *Artificial Intelligence: A Guide for Everyone*,
https://doi.org/10.1007/978-3-031-56713-1_11

multifaceted situations, whether they pertain to chess moves, medical diagnoses, or business strategies. By delving into the essence of these scenarios, AI systems can identify inherent challenges or obstacles that might impede progress. This capacity is akin to the critical thinking and decision-making abilities humans employ when faced with complex problems in various domains.

Fundamentally, AI reasoning relies on logical thinking, deduction, and inference, mirroring the cognitive processes employed by humans. Just as individuals evaluate information to make informed decisions, AI systems sift through data and leverage their knowledge to draw rational conclusions. This process significantly contributes to generating viable solutions or strategies to fulfill specific objectives. In essence, the reasoning and problem-solving facet of AI serves as a dynamic cog in the core mechanics of artificial intelligence, endowing machines with the acumen to tackle real-world challenges with efficiency and efficacy, with applications spanning from autonomous vehicles to healthcare diagnostics and beyond.

Probability Theory

Probability theory plays a fundamental and versatile role in reasoning and problem-solving across a wide array of domains, including artificial intelligence, statistics, and decision science. At its core, probability theory allows us to quantify and manage uncertainty, which is a pervasive aspect of real-world problems. In artificial intelligence, probabilistic reasoning techniques are instrumental in situations where uncertainty exists in data, models, or outcomes. For example, Bayesian networks, a prominent application of probability theory in AI, enable the representation of uncertain relationships between variables, making them valuable for tasks like medical diagnosis, fraud detection, and natural language understanding. These networks use probability distributions to model dependencies between variables and facilitate reasoning under uncertainty, offering a principled way to combine prior knowledge and observed data to make informed decisions.

Probability theory is vital in decision-making processes, particularly when faced with multiple choices and outcomes. Decision theory, which builds upon probabilistic foundations, assists in making optimal choices in situations with risks and uncertainties. In the context of reinforcement learning, a subfield of AI, agents learn to navigate complex environments and optimize their decision-making by assessing the probabilities of different actions leading to various rewards. Probability theory is essential in defining the uncertainty in these reward estimates and determining the most favorable actions.

In essence, probability theory equips AI systems and problem-solving methodologies with the capacity to handle real-world complexities, evaluate risks, and devise rational strategies that maximize desired outcomes while considering the uncertainties that may exist in the environment. This profound and versatile application of probability theory underscores its significance in enabling reasoning and problem-solving in AI and numerous other fields.

Economic Principles in AI

The application of economic principles to the field of AI represents a promising avenue for addressing the challenges posed by incomplete or uncertain information. Firstly, economic frameworks offer valuable insights into decision-making processes within AI systems. By incorporating concepts like utility theory and rational choice modeling, AI agents can make more informed choices when confronted with imperfect data, maximizing expected outcomes and adaptability. Secondly, these economic concepts are instrumental in optimizing resource allocation in AI applications. By drawing on principles of efficient resource allocation from economics, AI systems can ensure the judicious distribution of limited resources like computational power, budget, or attention, thereby improving overall system performance and resource utilization.

Concepts like utility theory, which quantifies preferences and outcomes, empower AI agents to make informed decisions that maximize expected utility, even in the face of imperfect information. In AI, this can be used in recommendation systems to predict which items a user might prefer based on their past behavior and preferences.

Furthermore, the application of game theory, another branch of economics, is used to model strategic interactions where multiple decision-makers (agents) make choices with uncertain outcomes. It enables AI systems to model and strategize in complex, dynamic environments, facilitating negotiation, cooperation, and competition among multiple agents. Game theory is employed in AI for scenarios like designing automated auctions, where agents bid on items with uncertain values.

Economic principles can be applied to optimize the allocation of limited resources, such as scheduling tasks on a computer cluster or routing deliveries for a logistics company. These optimizations often consider uncertain factors like varying demand or travel times.

Concepts Used by AI

In essence, AI leverages fundamental principles derived from probability theory and economics to facilitate the process of decision-making and forecasting in scenarios characterized by ambiguity or information gaps. These foundational concepts empower AI systems to enhance their decision-making capabilities by systematically considering and managing uncertainties, thereby enabling them to navigate the intricate dynamics of real-world situations.

Whether applied in the domain of medical diagnosis to offer precise and probabilistic insights into patients' conditions, utilized for fine-tuning financial planning strategies by accounting for market volatility and future uncertainties, or integrated into autonomous vehicles to facilitate safe and adaptive navigation through dynamic environments, these principles serve as the bedrock upon which AI systems operate, ensuring robust and reliable performance in the face of the world's inherent unpredictability.

Knowledge Representation

Overview

Knowledge representation is a foundational and pivotal component in AI, providing the means to capture, structure, and utilize information effectively. It is the process of encoding real-world knowledge, facts, and concepts into a format that machines can comprehend, manipulate, and reason with. Knowledge representation techniques are diverse and range from symbolic representations using logic, semantic networks, and ontologies to probabilistic models and vector-based embeddings.

The primary objective of knowledge representation in AI is to facilitate the storage, retrieval, and utilization of information for various cognitive tasks, including problem-solving, decision-making, natural language understanding, and machine learning, ultimately enabling intelligent systems to mimic human-like cognitive processes by harnessing the wealth of human knowledge and expertise.

Process

Knowledge representation in AI is a fundamental process that underpins the development of intelligent systems. It bridges human knowledge and machine understanding, allowing computers to manipulate and reason with complex information. It offers a structured means of encoding real-world knowledge and relationships, making it accessible for intelligent systems to process, query, and manipulate.

At its core, knowledge representation in AI involves encoding real-world knowledge, facts, concepts, and the relationships between them into a format that machines can comprehend. This encoding process is essential for allowing AI systems to perform tasks that require understanding and reasoning about the world, such as natural language understanding, problem-solving, decision-making, and learning. Effective knowledge representation forms the basis for building intelligent systems capable of mimicking human-like cognitive processes.

Knowledge Representation and Reasoning

Knowledge representation and reasoning (KRR) in AI is all about how to store, organize, and use information in a way that machines can understand and make decisions based on it, just like humans do. AI researchers have aimed to develop methods for computers to represent and store knowledge in a way that could be used for reasoning and decision-making.

Knowledge representation is fundamental to the study of the mind. In artificial intelligence, it describes the representation of knowledge. In simple terms, it is a study of how an intelligent agent's beliefs, intentions, and judgments can be expressed suitably for automated reasoning. A key purpose of knowledge representation includes modeling intelligent behavior for an agent.

Knowledge representation describes how we can represent knowledge in AI. It is not just storing data in some database—it also enables an intelligent machine to learn from that knowledge and experiences so that it can behave intelligently like a human. The types of knowledge that need to be represented in AI systems include objects, events, performance, metaknowledge (knowledge about what we know), facts, and knowledge base. The central component of knowledge-based agents is the knowledge base.

Crucial Role of KRR

Knowledge representation and reasoning in AI contribute to the intelligent behavior of agents. It represents information about the real world so that a computer can understand and utilize this knowledge to solve complex real-world problems, such as diagnosing a medical condition or communicating with humans in natural language.

Knowledge representation and knowledge engineering allow AI programs to answer questions intelligently and make deductions about real-world facts. Humans excel at understanding, reasoning, and interpreting knowledge. Based on that knowledge, they perform various actions in the real world. The way machines perform these comes under knowledge representation and reasoning.

In knowledge-representation algorithms, AI agents tend to think and contribute to making decisions. With the aid of such complex thinking, they are capable of solving complex problems in real-world scenarios, which are difficult and time-consuming for human beings to interpret.

Knowledge Representation Techniques

Several techniques are employed in knowledge representation, each with its strengths and suitability for different types of knowledge. Symbolic representation relies on formal logic and uses symbols and rules to represent knowledge. While effective for explicit, well-defined knowledge, it may struggle with uncertainty and complex relationships. On the other hand, semantic networks structure knowledge as interconnected nodes, with each node representing a concept and links depicting relationships. This approach is particularly suitable for organizing knowledge hierarchically.

Frames provide a structured framework for organizing knowledge, where each frame represents an object, concept, or situation and contains slots for attributes and values. They are useful for representing structured information. Logic-based representation uses logical statements, like those in mathematics, to express facts and relationships. For example, "All birds can fly" can be represented as a logical statement: "For all X, if X is a bird, then X can fly."

Ontologies formally represent knowledge, describing concepts, relationships, and constraints. They play a crucial role in knowledge engineering and are the backbone of the Semantic Web, which is an extension of the World Wide Web that aims to enhance the ability of machines to understand and process the content of web pages and data.

In contrast, probabilistic models use probability theory to represent uncertainty and capture complex relationships, making them ideal for areas like natural language processing, machine learning, and decision-making under uncertainty. Modern techniques, such as Word Embeddings or Graph Embeddings, represent knowledge in vector spaces, which are highly effective for machine learning applications, including text analysis and recommendation systems.

The choice of a knowledge-representation technique depends on the nature of the knowledge and the specific AI application. Effective knowledge representation is fundamental to the development of intelligent machines and the advancement of artificial intelligence as a whole. It empowers AI systems to perform cognitive tasks and provides a basis for mimicking human-like reasoning and comprehension, ultimately driving the field of AI forward.

Knowledge Reasoning Process

After the computer has knowledge represented, it can use it to make decisions, answer questions, or solve problems. This is where knowledge reasoning comes in, which is described in the following sections.

Deductive Reasoning

Deductive reasoning is a logical process that involves drawing specific conclusions from general principles or premises. In the given example, the process begins with two premises: "All birds can fly" and "Robins are birds." Through deductive reasoning, one can derive a specific conclusion: "Robins can fly." This conclusion is reached by applying the general rule that all birds have the ability to fly to the specific case of robins, which have been categorized as birds. Deductive reasoning operates on the principle that if the premises are true and the logical rules are valid, the conclusion must also be true. It is a method of reasoning commonly used in mathematics, philosophy, and formal logic, providing a systematic way to ensure the validity of conclusions drawn from established principles.

Inductive Reasoning

Inductive reasoning is a cognitive process where generalizations are derived from specific observations. When every observed instance of robins demonstrates the ability to fly, the general conclusion is drawn that "All robins can fly." This method extrapolates from particular cases to establish broader principles but is accompanied by a degree of uncertainty, as the generalization may not hold true in all circumstances. The inherent probabilistic nature of inductive reasoning reflects its reliance on observed patterns to infer overarching principles, making it a fundamental aspect of human and artificial intelligence alike.

Abductive Reasoning

Abductive reasoning is a form of inference that involves making educated guesses or forming hypotheses to provide plausible explanations for observed phenomena. In the context of seeing a robin in the sky, abductive reasoning would lead to the hypothesis that "The robin is flying because robins, as a species, are known to possess the ability to fly." This type of reasoning aims to identify the most likely explanation or cause behind an observation based on existing knowledge, patterns, and contextual information. Abductive reasoning is particularly useful in situations where there may be multiple plausible explanations, and it helps in generating hypotheses that align with observed data.

Probabilistic Reasoning

Probabilistic reasoning, similar to abductive reasoning, involves making educated guesses to explain observations. If a data point exhibits a certain pattern, it might be hypothesized that "The data point follows this pattern with high probability based on historical trends and probabilistic models." This approach entails leveraging statistical probabilities and data patterns to infer likely explanations for observed phenomena, providing a probabilistic framework for reasoning and decision-making in uncertain or complex situations.

Temporal Reasoning

Temporal reasoning in AI involves comprehending and analyzing how facts evolve or change over time. For example, by integrating information such as "John was born in 1980" with the current date "It's currently 2023," AI can deduce that "John is 43 years old." This capability is crucial for AI systems to make accurate predictions, adapt to dynamic scenarios, and understand the temporal aspects inherent in various datasets.

In AI systems, combining knowledge representation with various forms of reasoning allows machines to process information, draw conclusions, solve problems, and make informed decisions. This is crucial for tasks like natural language understanding, medical diagnosis, autonomous driving, and many other applications.

Planning

Overview

Planning in AI refers to the process of defining a sequence of actions or strategies to achieve specific goals or objectives within a given environment. It is a fundamental component of artificial intelligence that involves the generation and execution of a series of actions to bring about desired outcomes, even in complex and uncertain situations.

AI planning is used in various domains, such as robotics, autonomous systems, logistics, and game playing. It relies on algorithms and techniques to navigate decision spaces efficiently and determine the most effective course of action. The objective of AI planning is to create intelligent agents capable of making informed decisions and solving problems, allowing them to adapt and excel in dynamic and ever-changing environments.

Process

The planning process in AI is an intricate journey involving several vital components that allow intelligent agents to set goals, understand their environment, select actions, and navigate toward desired objectives. It initiates with goal formulation, explicitly defining the objectives an AI system needs to accomplish. Clarity in this stage is crucial for ensuring the entire planning process is built on a strong foundation.

Subsequently, the state-representation phase involves encoding information about the environment, such as object positions and relevant conditions, providing the AI agent with a comprehensive understanding of the world it interacts with. The planning process then necessitates knowledge of available actions, each with specified preconditions and effects, forming the agent's toolkit for effecting changes in the environment.

Search and planning algorithms are at the heart of the process, enabling the exploration of possible paths to reach the goals. These algorithms traverse the space of potential states and actions, evaluating their feasibility and adapting optimally to both straightforward and complex problem scenarios. After planning, the AI system moves into execution, systematically performing actions and adapting to changing circumstances as needed. Finally, the iterative nature of planning is vital, as the AI system learns and adapts its strategies based on feedback received during execution, allowing it to enhance its planning and decision-making capabilities for future tasks.

In essence, the planning process is the cornerstone of AI, and it functions as a dynamic cycle where AI systems repeatedly define goals, construct state representations, deploy search and planning algorithms, execute plans, and adapt strategies based on feedback. This process is applicable across diverse fields, from robotics and autonomous systems to logistics and scheduling. While it presents opportunities for innovation and advancement, the planning process also poses challenges, demanding effective coordination of multiple components and considerations. It remains a vibrant field, continuously evolving to meet the demands of increasingly complex AI applications.

AI Planning

Planning in AI is about determining the best sequence of actions to achieve a goal in a given environment. It involves modeling the world, searching for a plan, and often making use of heuristics and other optimization techniques to find efficient solutions. The following steps provide a more detailed view of what is involved in AI planning.

Goal Definition

Planning starts with defining a goal or objective. The AI system needs to know what it is trying to achieve, which could be anything from finding the shortest route in a navigation system to assembling a product in a factory.

State and Action Space

In an AI system, the representation of the world involves creating a framework with two essential components: a set of potential states that correspond to various situations or configurations and a set of possible actions that depict the agent's capacity to transition from one state to another. This model helps the AI agent understand and navigate its environment by defining the different scenarios it can encounter and the actions at its disposal to move between these scenarios effectively. By continually assessing and choosing actions based on this world model, AI systems can make informed decisions and adapt their behavior according to the dynamic conditions they encounter.

Search Algorithms

Once the AI system has a grasp of the potential states and available actions within its environment, it employs search algorithms to systematically investigate various sequences of actions to achieve a specific goal. These algorithms are designed to seek the most efficient or optimal path that leads from the current state to the desired goal state. Through this process, AI systems can evaluate multiple action sequences, assess their outcomes, and select the one that best aligns with the criteria of efficiency and effectiveness in reaching the intended objective.

Heuristics

In AI planning, the efficiency of the search process is often enhanced through the application of heuristics, which are essentially rules of thumb or educated estimates that guide the AI system in prioritizing actions. For instance, in the context of a navigation system, a heuristic might be as straightforward as "move toward the goal." These heuristics serve as valuable shortcuts by offering the AI system hints on which actions are more likely to lead to the desired outcome, thereby streamlining the decision-making process and improving overall efficiency.

Action Representation

In AI planning, it is imperative for planners to possess a comprehensive understanding of actions, including their feasibility, preconditions (the conditions that must be met before an action can be executed), and effects (how the state changes as a result of an action). This comprehensive knowledge is paramount in the planning process, allowing AI systems to make informed decisions and sequences of actions. By meticulously evaluating the preconditions and anticipating the effects of each action, AI planners can systematically work toward achieving their defined goals with precision and efficiency.

Partial-Order Planning

Partial-order planning, a problem-solving approach in AI, entails organizing actions based on dependencies and constraints, introducing flexibility in determining the optimal sequence of actions for goal achievement. This method is particularly advantageous in tasks where the order of actions is flexible, offering adaptability in varying conditions. Partial-order planners serve as valuable tools in the AI planning toolkit, effectively managing situations where action sequences can vary while still achieving the desired end state. This adaptability makes them versatile and efficient in addressing uncertainties and dynamic environments in problem-solving and goal attainment within AI systems.

Hierarchical Planning

Hierarchical planning is a strategy used to tackle complex tasks by breaking them down into more manageable subtasks. This approach involves structuring actions into a hierarchical framework, simplifying the management and resolution of extensive planning problems. By organizing actions into layers or levels of priority and dependency, hierarchical planning ensures a systematic and efficient method for accomplishing intricate objectives.

Plan Execution

Once a plan is generated, the execution phase involves implementing it in the real world. AI systems may encounter unexpected circumstances during execution, requiring them to adapt and revise their plans on the fly. This dynamic capability allows AI to respond effectively to changing environments and achieve its intended goals.

Applications

Planning is used in various AI applications, including robotics (where robots plan their movements), logistics (for optimizing supply-chain operations), game playing (to make strategic decisions), and more.

Automated Planning Competitions

The field of AI planning features competitions in which researchers develop and evaluate planning algorithms across a spectrum of problem domains. These competitions serve as a catalyst for advancing state-of-the-art planning techniques by fostering innovation, promoting healthy competition, and providing a benchmark for measuring planning system performance. Researchers and practitioners participate in these events to continually refine and improve planning algorithms, ultimately enhancing the capabilities of AI systems in various applications.

Learning

Overview

Artificial intelligence has long been dedicated to the goal of enabling machines to learn and improve their performance based on data and experiences rather than relying solely on explicit programming. Two fundamental components of AI research that contribute to this objective are machine learning and deep learning. Both machine learning and deep learning are aimed at developing algorithms and models that empower computer systems to learn and make predictions or decisions autonomously through exposure to data. In essence, they revolve around the idea of teaching computers to improve their performance automatically over time without the need for explicit programming.

Process

Machine learning, an essential subfield of AI, is inherently process-driven, revolving around the development of algorithms and models. These computational constructs serve as the backbone for computers to not only learn from data but also make predictions and decisions, thereby reducing their reliance on explicit programming. It is the dynamic and ongoing process of accumulating experience that stands at the core of machine learning's significance.

This dynamic nature is manifested in the relentless research efforts, driving its adaptation and practical applications across diverse industries, including healthcare, finance, and autonomous vehicles. The power of machine learning lies in its capacity to continuously refine its understanding and capabilities, resulting in the automation of intricate tasks, uncovering valuable insights from extensive datasets, and enhancing the process of decision-making. The iterative and learning-intensive process defines machine learning, propelling its evolution and expanding the boundaries of AI's potential applications. Through each data-driven interaction, machines become more adept, which is pivotal to the future growth of AI technology.

Machine Learning: Key Concepts

Data

As a data-driven approach, machine learning heavily depends on data as its primary source of information, drawing insights and patterns from vast and diverse datasets encompassing various forms, including text, images, numbers, audio, and any other structured or unstructured format.

Learning

Machine learning algorithms, driven by data, are designed to discern intricate patterns, correlations, and trends hidden within the vast amount of information they process, continuously fine-tuning their internal parameters based on the data they encounter.

Training

During the training phase, machine learning models are exposed to a dataset that contains examples of input data (features) and the correct output or target (labels). The model learns from this dataset to make predictions or decisions.

Prediction

After training, the machine learning model can take new, unseen data and make predictions or classifications based on what it has learned. For example, a trained model can predict whether an email is spam or not, based on its content and previous examples of spam and non-spam emails.

Types of Learning

Machine learning has two main types: supervised and unsupervised. They vary in their training methods and data conditions. Supervised and unsupervised learning each have unique strengths, making them suitable for different tasks. The third type, reinforcement learning, is based on an AI agent learning to make decisions based on the rewards it receives for actions in an environment.

Deep Learning

Deep learning is a subset of machine learning, which is particularly powerful for tasks like image and speech recognition. It has led to many breakthroughs in various AI applications.

Applications

Machine learning has a wide range of applications, including natural language processing (understanding and generating human language), computer vision (interpreting and processing images or videos), recommendation systems (suggesting products or content based on user behavior), and many more.

Evaluation

Machine learning models are subjected to rigorous evaluation, employing a suite of diverse metrics, including accuracy, precision, recall, and F1 score, tailored to the specific requirements of the tasks they address. Cross-validation techniques are frequently utilized to enhance the reliability of these assessments. This approach enables a more thorough exploration of the model's adaptability and performance across diverse data subsets, yielding valuable insights for refining and optimizing the model.

12

Additional Goals of AI Research

Natural Language Processing

Overview

Natural language processing is a pivotal field within the realm of AI, dedicated to bridging the communication gap between humans and machines. It empowers computers to not only understand but also effectively respond to human language, both in text and in speech, enriching the human–machine interaction with intuition and naturalness.

NLP revolves around a complex computational process applied to human languages, involving functions like searching, analyzing, comprehending, and information extraction from textual input. At its core, NLP is designed to make user interactions with technology more seamless, intuitive, and human-like, fostering a profound transformation in the way we communicate with and benefit from AI systems. NLP enables computers to communicate with humans in a way that feels like having a conversation with a real person, making technology more user-friendly and accessible.

NLP applications include voice-operated GPS, digital assistants, speech-to-text dictation software, customer service chatbots, and other consumer conveniences. NLP also powers language translation apps, helps search engines understand queries, and even checks grammar and spelling in word processing software. NLP is also playing a growing role in enterprise solutions that help streamline business operations, increase employee productivity, and simplify mission-critical business processes.

© The Author(s), under exclusive license to Springer Nature Switzerland AG 2024
A. Khan, *Artificial Intelligence: A Guide for Everyone*,
https://doi.org/10.1007/978-3-031-56713-1_12

Process

NLP stands at the forefront of AI applications, acting as the lynchpin between technology and human communication. In practical implementation, NLP takes center stage as a foundational component of AI systems, expertly designed to engage with humans through text or voice inputs. Its core function goes beyond the surface-level comprehension of human language, delving into the intricate world of natural language understanding. At its essence, NLP thrives on a multifaceted computational process that harmoniously bridges the gap between machines and humans. This intricate process is the heart of NLP, enabling the effective and intuitive interaction between technology and its users.

The comprehensive nature of NLP's process involves an array of sophisticated functions. It engages in the task of sifting through vast volumes of text and voice inputs, meticulously searching for patterns, identifying relationships, and comprehending contextual nuances. It does not stop at comprehension; NLP has the remarkable ability to extract valuable information from the data it encounters, making knowledge retrieval a seamless and efficient endeavor.

These capabilities collectively elevate the quality of information retrieval and knowledge extraction, significantly enhancing the user's experience and the utility of AI systems. NLP represents the culmination of years of research and development in the quest to replicate human-like communication in AI, serving as a testament to the ever-evolving landscape of technology that continues to reshape the way we interact with machines.

NLP Functionality

Understanding Language

NLP is instrumental in facilitating computer comprehension of spoken and written language, enabling machines to interpret words, sentences, and the contextual significance embedded within text or speech. This capability extends to recognizing semantic equivalences, as evidenced by NLP's ability to deduce that terms like "cat" and "kitty" convey identical meanings, fostering more nuanced and effective human–computer interactions.

Making Sense of Text

NLP excels at the rapid summarization of extensive textual content, resembling the function of having a computer read a book and provide concise summaries. Leveraging NLP's capabilities, users can efficiently extract key insights and pertinent information from voluminous texts, streamlining information retrieval and decision-making processes across diverse applications, from research to content curation.

Translation

NLP exhibits the capacity to seamlessly translate text from one language into another, bridging linguistic barriers and facilitating effective communication between individuals who speak different languages. This powerful translation capability finds extensive application in diverse domains, including international business, cross-cultural collaboration, and global information dissemination, greatly enhancing the accessibility and reach of information across linguistic divides.

Chatbots and Virtual Assistants

Virtual assistants like Siri, Alexa, or website chatbots leverage NLP to comprehend user inputs and deliver responses that emulate human conversation, enhancing user interaction and convenience. NLP's ability to understand and generate human-like responses in real-time plays a pivotal role in the development of these conversational AI systems, enabling them to provide efficient and engaging user experiences across a wide range of applications and platforms.

Sentiment Analysis

Through sentiment analysis, NLP empowers businesses to discern whether a review or customer feedback carries a positive or negative sentiment, offering invaluable insights into public perception of their products or services. This capability serves as a crucial tool for companies seeking to gauge customer sentiment, identify areas for improvement, and make data-driven decisions to enhance customer satisfaction and brand reputation.

Spam Filters

NLP functions are akin to an intelligent email filter, adept at distinguishing between spam and important emails, streamlining inbox management for users, and ensuring priority communication is not overlooked.

Perception

Overview

Perception is a foundational element in AI, allowing machines to comprehend and interact with the world. This critical component covers various sensory modalities like computer vision, speech recognition, and audio processing, enabling AI systems to understand visual, auditory, and textual information. Through sensory data processing, perception interprets the environment and supports tasks such as image recognition, language understanding, and autonomous navigation. It acts as the sensory gateway for AI systems, enabling informed decisions and effective navigation in complex real-world scenarios.

In AI applications, perception is essential, allowing machines to react like humans. Robotics, for instance, relies on perception to handle object manipulation, navigation, and related challenges. These subtasks encompass localization, which involves determining the machine's position; mapping, which entails understanding the surroundings; and motion planning, which focuses on charting the path to a destination.

Examples of perception applications encompass computer vision, machine hearing, machine touch, and machine smelling. Machine perception harnesses sensor data from sources like cameras, microphones, wireless signals, LIDAR, sonar, and radar. These capabilities find utility in applications like autonomous vehicles, robotics, computer vision, speech recognition, facial recognition, object identification, and virtual reality.

Process

Perception is the process by which sensory information from the real world is interpreted, acquired, selected, and organized. Human beings possess sensory receptors for touch, taste, smell, sight, and hearing, with information from these receptors being transmitted to the human brain for organization.

In AI, perception is the process of interpreting sensory inputs, which can encompass vision, sounds, smell, and touch. The primary aim is to equip computers with the ability to sense, learn, and interact akin to humans. It involves the collection and processing of data from various sources to comprehend and engage with the environment. In the context of AI, perception refers to the machine or computer system's capacity to sense and interpret real-world information, often through sensors or data input.

Machine perception is the process through which sensory data from the physical world is interpreted, acquired, selected, and organized to execute actions, similar to how humans do. This enables computers to incorporate sensory input alongside traditional computational methods, resulting in more accurate information gathering and presentation.

Key Process Steps

Sensors

Perception, a foundational process in the realm of artificial intelligence and robotics, frequently commences with the deployment of sensors capable of capturing and gathering data from the surrounding environment. These sensors can be cameras, microphones, touch sensors, GPS devices, or any other hardware that collects data.

Data Collection

Sensors serve as data-capturing instruments, collecting raw information that may manifest in various formats, including images, sounds, text, or numerical measurements, depending on the specific sensing modality. For example, a self-driving car may use cameras and LIDAR sensors to capture images and depth information from the road and surroundings.

Data Preprocessing

As the initial output of data-capturing instruments, raw sensor data often presents itself in a noisy and unstructured manner, replete with imperfections and extraneous information. Reprocessing techniques are meticulously applied, aiming to clean, filter, and organize the information into a more refined and usable format, to render this data suitable for analysis and application. For instance, in image recognition, preprocessing might involve resizing and enhancing images.

Feature Extraction

The next critical step involves the extraction of key features or distinctive characteristics from the data, enabling a more focused and insightful understanding of the information. In the context of image processing, for instance, this feature extraction process entails identifying and isolating salient elements such as edges, colors, and shapes, which are instrumental in facilitating the recognition and differentiation of objects within the visual data. These extracted features serve as the foundation for subsequent analysis and decision-making.

Pattern Recognition

AI algorithms, frequently encompassing machine learning or deep learning models, serve as the computational backbone for the critical task of recognizing intricate patterns or deriving meaningful insights from the features that have been extracted from the data. For example, in speech recognition, a model learns to convert audio waves into transcribed text.

Contextual Understanding

Perception goes beyond recognizing patterns—it involves understanding the context of the data. For instance, in natural language processing, understanding the meaning and sentiment of words in the context of a sentence is crucial.

Decision-Making

Once the AI system has successfully processed and comprehended the data it has perceived, it is equipped with the capacity to initiate informed decision-making processes or execute specific actions in response to its perception. For instance, an autonomous drone can use perception to navigate and avoid obstacles.

Feedback Loop

In numerous AI systems, the process of perception unfolds as a continuous and dynamic feedback loop where the system persistently engages with its surroundings. This ongoing cycle entails the perpetual sensing of the environment, the real-time processing of newly acquired data, the iterative refinement of the system's understanding, and the subsequent adaptation of its actions in response to the evolving input. This feedback-driven approach enhances the system's responsiveness. It ensures that it remains adaptable to changing conditions and complexities, making it well-suited for various applications, from autonomous vehicles and industrial automation to smart environments and healthcare technologies.

Motion and Manipulation

Overview

Motion and manipulation in AI represent the domain of technology and research focused on enabling machines to interact with the physical world through controlled movements and actions. The central objective is to develop algorithms and systems that empower machines to perceive their environment, make informed decisions, and carry out physical tasks with precision and dexterity. By connecting digital intelligence to the physical world, including motion and manipulation, AI has the potential to revolutionize industries, enhance human–machine collaboration, and make complex physical operations safer and more efficient.

The overarching goal of the motion and manipulation process is not only to enhance the efficiency and precision of physical tasks but also to promote the safe coexistence and collaboration between humans and machines, heralding an era of automation and AI-enabled physical operations that are both efficient and secure.

Motion planning is fundamental for a wide range of applications, including robotics, drones, warehouse logistics, autonomous vehicles, and industrial automation, where machines need to perform tasks that involve physical movements and interactions with their surroundings. These systems aim to improve the efficiency and accuracy of tasks that would be challenging or dangerous for humans to perform manually.

Process

The motion and manipulation process in AI is a multifaceted domain focused on empowering machines to execute a wide array of physical tasks and engage with the tangible world in a purposeful manner. These tasks range from intricate activities such as controlling robotic arms with precision to the seemingly simple yet vital function of grasping and handling objects in diverse environments. This process is an amalgamation of several advanced technologies, notably including computer vision, sensor data processing, and sophisticated motion control mechanisms. These components work in tandem to equip machines with the capability to interact with and adapt to their surroundings, laying the foundation for the seamless integration of AI into various physical tasks and operations.

At the core of motion and manipulation in AI is the development of algorithms that facilitate machines in making informed decisions regarding their physical actions. These algorithms allow machines to perceive their surroundings through sensors and camera systems, process the collected data to generate actionable insights, and subsequently execute precise and context-aware movements. This concept extends its influence across a spectrum of real-world applications, spanning industrial automation, where robots efficiently manage complex manufacturing processes, to the field of autonomous vehicles, where machines must navigate dynamically changing environments safely and effectively.

Motion Planning in AI

Motion planning is the process of breaking down a movement task into individual steps or a sequence of steps. From experience, robots can learn how to move efficiently despite the presence of friction and gear slippage. Motion planning can enable a collision-free path for a moving entity through an obstacle-filled environment, from an initial placement to a goal placement. While motion planning is critical for manipulation, it also plays a critical role in a wide variety of other application areas.

Manipulation manifests itself in many ways. Humans and robots can manipulate objects by carrying, holding, squeezing, pulling, pushing, throwing, and dropping them. Robots can do so with a single arm and hand or with both arms and hands. AI can respond to and manipulate uncommon changes in real-time.

In the AI context, motion planning refers to the process of generating a sequence of actions or movements for an autonomous agent, such as a robot or a self-driving car, to navigate through an environment while avoiding obstacles and reaching a specific goal or destination safely and efficiently. It is a crucial component of robotics and autonomous systems.

Key Aspects of Motion Planning

Environment Modeling

The AI system first needs to create a model of the environment to perform motion planning. This model typically includes information about obstacles, the agent's initial position, the goal, and any other relevant details.

State Space

The environment is often represented as a state space, where each state represents a possible configuration of the agent and its surroundings. For example, in a 2D grid-based environment, each cell in the grid can be a state.

Pathfinding

The goal of motion planning is to find a safe and feasible path from the agent's current state to the goal state. This path should avoid collisions with obstacles and obey any constraints, such as the agent's physical limitations.

Obstacle Avoidance

A significant part of motion planning involves avoiding obstacles. AI algorithms use various techniques to detect and avoid collisions, such as potential field methods, probabilistic roadmaps, or rapidly exploring random trees (RRTs).

Cost Functions

Motion planning, a fundamental aspect of AI-driven navigation, goes beyond merely identifying a path, as it frequently involves optimizing various crucial factors such as speed, energy efficiency, and safety. This process entails the utilization of cost functions, which assign respective costs to different potential paths, enabling the AI system to systematically seek out the path that minimizes the overall cost.

Dynamic Environments

Motion planning becomes notably more complex and demanding when applied in dynamic environments characterized by unpredictable changes in obstacles and conditions. In such scenarios, AI systems face the arduous task of initially charting a safe and efficient course and continually monitoring and recalibrating their plans to adapt in real-time to the evolving environment.

Real-Time Planning

Some applications, like autonomous vehicles, require real-time motion planning, where the system must make decisions quickly to respond to changing situations on the fly. In scenarios where robots interact with humans, motion planning also considers social and ethical aspects, ensuring that the robot's movements are safe and comfortable for people.

Learning-Based Approaches

Recent advancements in AI, including deep reinforcement learning, have been applied to motion planning. These methods allow agents to learn and improve their planning strategies through trial and error.

General Intelligence

Overview

General intelligence in AI is a multifaceted concept that represents the pursuit of creating machines with human-like cognitive abilities and a broad understanding of various tasks and domains. Unlike narrow AI, which excels in specific applications, general AI aims to imbue machines with the capacity to learn, reason, and adapt across a wide range of tasks and environments, much like the general intelligence exhibited by humans.

Achieving this level of AI involves the development of versatile algorithms and models that can not only solve problems but also generalize their knowledge, enabling machines to apply what they have learned in one domain to new, unfamiliar situations. General intelligence in AI remains a long-term aspiration, with researchers continually exploring novel techniques and methodologies to bring machines closer to human-level cognitive capabilities, which has the potential to transform industries, enhance automation, and broaden the scope of AI's applications.

Ultimate Goal

The ultimate goal of AI research is to create systems that possess general intelligence, often referred to as AGI or general intelligence. While many AI systems are designed for specific tasks, the ultimate goal for some researchers is to create AGI, which would possess human-like intelligence and the ability to perform a wide range of cognitive tasks.

The AGI concept aims to create machines or computer systems that possess human-like cognitive abilities and the capacity to perform a wide range of tasks without specific programming for each task. Unlike specialized AI systems designed for specific tasks (such as playing chess or recognizing faces), AGI seeks to develop machines that can understand, learn, and adapt across a wide range of tasks and domains, much like the general intelligence exhibited by humans. In essence, it is about building AI that can think, reason, learn, and problem-solve in a flexible and versatile manner, just like humans do.

Unlocking AGI's Potential: A Leap in AI Capabilities

Imagine a computer system that can not only excel at tasks it was explicitly programmed for but also learn and apply its knowledge to entirely new and diverse challenges, just as a human can switch from solving math problems to playing a musical instrument to engaging in a conversation about history. AGI represents the aspiration to create AI that can understand the world, learn from experiences, make sense of complex information, and exhibit a level of adaptability and problem-solving that approaches or even surpasses human capabilities. Achieving AGI is a profound goal in AI research because it would open up a world of possibilities for creating highly intelligent machines capable of tackling a wide array of real-world problems.

13

Machine Learning

Fundamentals

Concept

Machine learning is one of the most dynamic fields in advanced computer technology. It is based on the concept that systems/machines can learn from data, recognize patterns, and improve performance or make decisions with little or no human input. A computer "learns" when its software is able to successfully predict and react to unfolding scenarios based on previous outcomes.

Functionality

Machine learning enables computer systems, programs, or applications to learn automatically and, subsequently, develop better results based on experience without being programmed to do so. It can continuously learn from and make predictions based on data. Machine learning can make adjustments without being specifically programmed to do so. It helps find patterns in data, uncover insights, and improve the desired results based on past experience.

Machine learning can find hidden correlations in various data. With this information, it can create a predictive model that is able to pinpoint future issues, such as the failure of a car component like the brakes. It may even be able to predict the failure timing, which will provide an alert to the owner to replace the at-risk component.

A. Khan, *Artificial Intelligence: A Guide for Everyone*,
https://doi.org/10.1007/978-3-031-56713-1_13

Applications

Machine learning, a dynamic and ever-evolving AI field, has unveiled a plethora of possibilities that have redefined how we interact with technology. Its applications span across numerous domains, transforming our daily lives, businesses, and industries in multifaceted ways. From recognizing images and videos to facilitating medical diagnoses, from enhancing financial forecasting to enabling self-driving cars, machine learning has woven itself into the fabric of our modern world. In this section, we will delve into some of the remarkable capabilities of machine learning and how they are reshaping the way we perceive and engage with technology.

One of the most remarkable feats of machine learning lies in its ability to recognize patterns, objects, and faces in images and videos. Facial recognition systems and security measures have harnessed this technology to enhance identification and authentication processes. The advent of autonomous systems, such as self-driving cars, drones, and robots, is largely made possible by machine learning. These machines utilize advanced algorithms to perceive their surroundings, make decisions, and navigate safely through complex environments, offering us a glimpse into a future with increased automation.

Natural language processing, a domain of machine learning, has unlocked the potential for computers to understand and generate human language. NLP techniques are behind the chatbots that assist us in online interactions, language translation services that bridge linguistic divides, sentiment analysis tools that gauge public opinion, and speech recognition systems that transcribe our words into text.

Machine learning excels in creating recommendation systems that analyze user behavior and preferences to offer personalized suggestions. Whether it is movies on streaming platforms or products on e-commerce websites, these systems have revolutionized how we discover content and make choices.

In the field of healthcare, machine learning plays a pivotal role in medical diagnosis. By analyzing medical images like X-rays, MRIs, and CT scans, machine learning models assist in detecting diseases and predicting patient outcomes. This technology is not just revolutionizing diagnosis but also contributing to the formulation of effective treatment plans. Machine learning also assists in predicting disease outbreaks, discovering new drugs, and monitoring patient well-being by analyzing health data for trends and risks.

Financial markets benefit from machine learning's prowess in predicting stock prices, detecting fraudulent activities, and managing risk. Machine learning algorithms increasingly guide investment decisions and portfolio management.

Anomaly detection, another hallmark of machine learning, identifies unusual patterns or outliers in data, which is invaluable for fraud detection and network security. Manufacturing industries benefit from machine learning's quality control capabilities, ensuring consistent product quality by inspecting for defects.

Machine learning has ushered in a revolutionary era for language translation services, enabling precise and fluent translation of both written text and spoken words across a multitude of languages. This technological advancement has not only broken-down language barriers but also fostered global communication and collaboration with unprecedented accuracy and ease.

In the world of gaming, machine learning algorithms have surpassed human champions in strategic games like chess and Go. They employ reinforcement learning to continually refine their strategies.

Machine learning is also the driving force behind personalized experiences on websites and apps. Content recommendations, social media feeds, and news suggestions are customized for individual users, enhancing their online interactions. Customer support has seen a transformation with the introduction of chatbots and virtual assistants that employ machine learning to provide automated assistance, answering questions and resolving issues.

Machine learning significantly enhances environmental monitoring by harnessing its capacity to analyze extensive climate data, accurately predict forthcoming weather patterns, and vigilantly oversee shifts and alterations within the environment. This application of machine learning empowers proactive environmental management, aiding in climate research and the mitigation of potential natural disasters.

Machine learning models exhibit the remarkable ability to generate text that closely resembles human language, fostering innovation in diverse fields such as content generation, where they can craft compelling and contextually relevant written material. Moreover, these models can extend their creative prowess to music composition, producing melodies and harmonies that resonate with the artistry of human composers.

In conclusion, the influence of machine learning extends far and wide, shaping industries, enhancing daily life, and redefining our interactions with technology. The diverse applications of machine learning, from image recognition to healthcare, highlight its transformative power and potential to usher in a future characterized by innovation and automation. As machine learning advances, it will undoubtedly uncover new possibilities and reshape the landscape of our technological world.

Types of Machine Learning

Three Methods

Machine learning encompasses various methods and techniques. The three fundamental methods of machine learning are as follows:

- Unsupervised learning
- Supervised learning
- Reinforced learning

These three methods represent broad categories within machine learning, but the field is vast, and there are many other specialized techniques and sub-fields, such as semi-supervised learning, deep learning, and transfer learning, each suited to different types of problems and data. The method to be used depends on the problem being solved and the nature of the data.

Unsupervised Learning

Objective

Unsupervised learning is a type of machine learning which finds patterns in a stream of input. Its primary objective is to extract useful information and gain insights from data, often for purposes like data clustering or dimensionality reduction. It involves training algorithms on unlabeled data, where the algorithm must find patterns, group similar data points, or discover hidden structures within the data.

These algorithms discover hidden patterns or data groupings without the need for human intervention. Since they are trained on unlabeled data, it means that the data used for training does not have predefined target labels or categories. Instead, the algorithm tries to find patterns, structures, or relationships within the data, such as clustering similar data points together or reducing the dimensionality of the data on its own.

Characteristics of Unsupervised Machine Learning

No Target Labels

Unlike supervised learning, where the algorithm is trained on labeled data (input data with corresponding output labels), unsupervised learning cannot access target labels. This means the algorithm does not know the correct answers in advance.

Clustering

A common task in unsupervised learning is clustering, where the algorithm groups similar data points together into clusters or categories. For example, in customer segmentation, the goal might be to group customers based on their purchase behavior, but without knowing in advance how many segments there should be or what those segments represent.

Dimensionality Reduction

Unsupervised learning can also be used for dimensionality reduction. This means reducing the number of features or variables in a dataset while preserving its essential characteristics. Principal component analysis (PCA) is a popular technique for dimensionality reduction.

Anomaly Detection

Another application is anomaly detection, where the algorithm identifies data points that are significantly different from most of the data. This can be useful for detecting fraud, network intrusions, or defects in manufacturing.

Feature Learning

Unsupervised learning can be used for feature learning, where the algorithm learns useful representations or features from the data. These representations can then be used as input for other machine learning tasks, such as supervised learning.

Data Exploration

Unsupervised learning can also be used for exploratory data analysis. It helps researchers and data scientists understand the underlying structure of the data, discover hidden patterns, and gain insights into the dataset.

Algorithms

Some common unsupervised learning algorithms include k-means clustering, hierarchical clustering, principal component analysis, independent component analysis (ICA), t-distributed stochastic neighbor embedding (t-SNE), and generative models (autoencoders), among others.

Application

The ability of unsupervised learning to discover similarities and differences in information makes it the ideal solution for exploratory data analysis, cross-selling strategies, customer segmentation, and image recognition. It is valuable for situations where labeled data may not be available or when it is desired to explore and understand the data structure before applying other machine learning techniques.

Unsupervised learning is widely used in fields like data analysis, natural language processing, and image processing, where finding patterns and structures within data can provide valuable insights and facilitate further analysis.

Supervised Learning

Objective

The primary goal of supervised learning, also known as supervised machine learning, is to learn a mapping or relationship between inputs and outputs so that the algorithm can make predictions or classifications on new, unseen data. In this type of learning, the algorithm is trained on a labeled dataset, where each input example is paired with the correct output or target variable. The training dataset, which teaches models to yield the desired output, includes inputs and correct outputs, which allows the model to learn over time.

Supervised learning uses labeled datasets, meaning the training dataset includes input data and their corresponding output labels or target values. This means the input data is paired with the correct output to train algorithms to accurately classify data or predict outcomes. These algorithms learn from labeled data, and as input data is fed into the model, it adjusts its weights until the model has been fitted appropriately, which occurs as part of the cross-validation process.

Characteristics of Supervised Machine Learning

Labeled Data

In supervised learning, there exists a dataset in which each example consists of input data and the correct answer or label. For instance, in a spam email detection task, each email is labeled as either "spam" or "not spam."

Prediction or Classification

Supervised learning can be used for both prediction and classification tasks. In prediction tasks, the algorithm predicts a continuous value, such as predicting the price of a house based on its features. In classification tasks, the algorithm assigns input data to predefined categories or classes, such as classifying images of animals into "dog," "cat," or "horse."

Model Training

During the training phase, the algorithm learns to make predictions or classifications by analyzing patterns and relationships in the labeled data. It adjusts its internal parameters based on the training data to minimize the difference between its predictions and the actual labels.

Evaluation

After training, the performance of the supervised learning model is evaluated on a separate dataset that was not used during training. Depending on the task, common evaluation metrics include accuracy, precision, recall, F1 score, and mean squared error.

Types of Supervised Learning

Algorithms use loss functions to measure their accuracy or performance. The loss function quantifies how far the algorithm's predictions are from the actual values in a supervised learning setting. The algorithm's objective is to minimize this loss function, adjusting until the error has been sufficiently minimized, which implies reducing the prediction error. This method has two main varieties: regression and classification.

Regression

Regression tasks involve machine learning algorithms predicting continuous outputs, such as estimating variables like temperature, stock prices, or individuals' ages, by analyzing the relationships within the data to provide valuable numeric predictions for a wide range of applications.

Classification

Classification tasks in machine learning entail the algorithm categorizing input data into distinct classes or categories, serving various purposes, from filtering email as spam or not, diagnosing the presence or absence of a disease, to analyzing sentiment and classifying it as positive, negative, or neutral. This process involves the algorithm utilizing pattern recognition and data analysis to make discrete and well-defined decisions within a multitude of real-world applications.

Algorithms

Algorithms in supervised learning can perform both regressions (predicting continuous values) and classification (categorizing data into classes), depending on the specific problem and data. While labeled data is typically associated with supervised learning, it is not the sole factor distinguishing between regression and classification. In supervised learning, algorithms can perform both regression and classification tasks, depending on the nature of the output variable. The key distinction between regression and classification is the type of output variable they predict, not solely the use of labeled data.

Application

Supervised learning algorithms include linear regression, logistic regression, decision trees, support vector machines, and neural networks, among others. Supervised learning includes tasks like image classification (identifying objects in images) and natural language processing (text classification, sentiment analysis).

Supervised learning helps organizations solve a variety of real-world problems at scale, such as classifying spam so that it can be segregated and sent to the appropriate mail folder. Supervised learning is used for tasks like classification and regression. It is widely used in various fields and applications, including natural language processing, computer vision, speech recognition, recommendation systems, and many others. It is a foundational technique in machine learning and is especially useful when there is a need to access labeled data to train and evaluate models.

Reinforcement Learning

Objective

Reinforcement learning is a type of machine learning where an AI agent learns to make sequences of actions or decisions by interacting with an environment in order to maximize a cumulative reward signal. The agent aims to learn a policy (a strategy) that leads to the best long-term outcomes by exploring different actions and receiving feedback through rewards or penalties.

Reinforcement learning is inspired by behavioral psychology, where learning is driven by rewards and punishments. In reinforcement learning, the agent aims to maximize a cumulative reward signal over time through a series of actions it takes in the environment. It learns from the consequences of its actions over time. It is often used in scenarios like training autonomous agents or game playing.

Characteristics of Reinforcement Learning

Agent

An agent, in the context of AI and machine learning, refers to a system that is actively engaged in the process of learning how to execute tasks within a given environment. This fundamental concept underpins the development and training of AI systems, enabling them to acquire skills and adapt their behavior to achieve specific objectives.

Environment

In the field of AI and machine learning, the term environment pertains to the external system or context in which an AI agent operates. This environment can manifest in various forms, encompassing virtual simulations within computer programs, real-world scenarios where robots traverse physical spaces or any context that provides the agent with opportunities to execute actions. The dynamic interaction between the agent and its environment plays a fundamental role in shaping the agent's learning process, enabling it to adapt and make informed decisions within the given context.

State

In AI terminology, a state refers to the prevailing conditions or arrangement of elements within the environment in which an agent is interacting. This state is the foundational data that guides the agent's decision-making process, influencing its actions as it navigates and operates within its surroundings.

Action

Action in AI parlance pertains to the array of potential moves or decisions accessible to an agent when situated within a particular state. Each action within this set may result in distinct consequences or outcomes, and the agent's primary objective is to make choices that lead to favorable results or align with its predefined goals. By intelligently selecting actions based on its evaluation of potential outcomes, the agent can navigate its environment and achieve its intended objectives.

Reward

Reward in the context of AI reinforcement learning represents a numeric signal that the agent obtains from its environment following the execution of an action within a specific state. These rewards act as a form of feedback, conveying to the agent whether the action it took was advantageous or detrimental in achieving its goals. The core aim for the agent is to strategize and make decisions that lead to the accumulation of the highest possible cumulative reward over a sequence of actions and states, a fundamental principle guiding the learning process in reinforcement learning. By learning from its experiences and the rewards received, the agent can iteratively refine its decision-making strategies to attain optimal outcomes.

Policy

In AI reinforcement learning, a policy refers to a systematic strategy or predefined rules that guide the agent in making decisions across various states within its environment. The overarching objective is to facilitate the acquisition of an optimal policy that consistently yields the highest expected cumulative reward over time. The development of such policy hinges on the agent's ability to learn and adapt its decision-making processes through trial and error, gradually converging toward actions that maximize its long-term rewards. As the agent refines its policy, it enhances its capacity to navigate complex environments and achieve its specified objectives with greater efficiency and effectiveness.

Algorithms

Reinforcement learning algorithms use trial and error to learn the best policy for the agent. The agent explores different actions in various states, receives rewards or penalties, and updates its policy based on this feedback to improve its decision-making over time. Popular algorithms in reinforcement learning include Q-learning, deep Q-networks (DQN), and proximal policy optimization (PPO).

Categorization Methods

In reinforcement learning, classification and regression can play important roles in different aspects of the learning process. Classification helps the AI agent categorize its current situation, while regression helps it estimate the expected rewards for different actions, both of which are crucial for making good decisions in reinforcement learning.

In reinforcement learning, classification determines which category something belongs to. It learns by looking at different groups and then figures out where new things fit. It is often used to categorize or label the different states or situations that the AI agent can encounter in its environment. The program uses examples from several categories to learn and classify new inputs. For example, if AI is learning to play a video game, it may use classification to determine if it is in a safe or dangerous state. This categorization helps the agent make decisions based on the current situation. It can be viewed as the agent deciding whether it is in a "good" or "bad" situation and adjusting its actions accordingly.

Regression involves finding a rule that shows how things are connected. It helps AI predict how one thing changes when something else changes. It can come into play when AI needs to estimate or predict numerical values, especially when dealing with rewards. It produces a function that describes the relationship between inputs and outputs and predicts how the outputs should change as the inputs change. For instance, if AI is learning to control a robot, it might use regression to predict how much reward it will receive for different actions. This prediction helps the agent choose actions that are expected to lead to higher rewards. In simple terms, regression helps the agent figure out how much it can expect to gain from its actions.

Application

Reinforcement learning finds versatile applications across numerous domains, including game playing, robotics, autonomous vehicles, recommendation systems, and beyond. In these diverse arenas, reinforcement learning comes into play when agents must acquire the ability to make sequential decisions in environments characterized by dynamism and uncertainty. Whether it is training AI agents to master complex games, navigate physical spaces, optimize vehicle operations, or provide personalized recommendations, reinforcement learning serves as a pivotal paradigm for imbuing machines with the adaptability and intelligence required to excel in real-world scenarios.

14

Machine Learning Development Process

Building a Machine Learning System

Foundational Elements

To develop a comprehensive machine learning system, developers build algorithms that can help improve perception, knowledge, thinking, or actions based on experience or data. They use computer science, statistics, psychology, neuroscience, economics, and control theory for this. Developing a comprehensive machine learning system involves several key components and steps.

The following are the foundational elements that lay the groundwork for the creation of an effective machine learning system. They encompass crucial elements, from data and algorithms to computational power and ethical considerations, ensuring a solid foundation for the development of AI systems.

Problem Definition

The first critical step in developing a machine learning system is to clearly define the problem it aims to solve. This involves articulating the problem's scope, goals, and expected outcomes, providing a solid foundation for the project.

A. Khan, *Artificial Intelligence: A Guide for Everyone*,
https://doi.org/10.1007/978-3-031-56713-1_14

Data Collection

This involves acquiring high-quality and relevant data essential for effectively training and testing the machine learning model. The system's performance is profoundly influenced by the quantity and quality of the data, making data collection a crucial aspect of the project. Careful consideration and curation of the dataset ensures that the model can learn and generalize patterns accurately.

Data Preprocessing

This involves preparing the data for analysis by cleaning it, handling missing values, and converting it into a suitable format. Data preprocessing may also involve feature engineering, which is selecting or creating informative features that help the model learn.

Algorithm Selection

This involves choosing the appropriate machine learning algorithm(s) based on the nature of the problem. Common algorithms include decision trees, support vector machines, neural networks, and more specialized models like convolutional neural networks for image data and recurrent neural networks for sequential data.

Model Training

To train the machine learning model effectively, the training dataset, which is a designated portion of the data available to the AI system, is prepared. Through this training process, the model acquires the ability to identify intricate patterns within the data and, consequently, can make informed predictions or classifications based on the provided input information. This training phase plays a crucial role in equipping the model with the necessary knowledge and understanding to perform well in practical applications.

Hyperparameter Tuning

The model's hyperparameters, settings that control the learning process, are fine-tuned to optimize its performance. This often involves experimenting with different parameter values and techniques.

Validation and Testing

The model's performance is assessed on a separate dataset (the validation set) to ensure it generalizes well to unseen data. Rigorous testing is conducted using a test dataset to evaluate its real-world performance.

Deployment

The trained model is implemented seamlessly within the target environment or application, ensuring its effective integration into the existing software infrastructure or systems. This process might entail creating APIs or interfaces that allow the model to interact with other components, guaranteeing its smooth functionality.

Monitoring and Maintenance

Continuous vigilance is maintained over the model's real-world performance to ensure its effectiveness. Over time, models can lose accuracy as data distributions shift, which necessitates regular updates and maintenance to adapt to evolving circumstances. This ongoing monitoring and fine-tuning guarantee that the machine learning system remains a reliable and valuable asset in achieving the objectives.

Scalability

When designing a machine learning system, it is crucial to consider its scalability to accommodate extensive data volumes and meet growing user demands. Achieving scalability might entail optimizing the model for efficiency or leveraging distributed computing resources to ensure the system can handle increased workloads without compromising performance.

Data Security and Privacy

It is paramount to ensure the system adheres to stringent data protection regulations, safeguarding user privacy and sensitive information. Therefore, robust security measures are implemented to fortify the system against potential threats and breaches, instilling trust and confidence in users.

Ethical Considerations

Implementing ethical machine learning involves addressing bias, fairness, and transparency, through various strategies. Avoiding unintended discrimination and providing clear explanations for model decisions are important.

Documentation and Reporting

This element necessities the maintenance of thorough documentation of the work, including data sources, model details, and any updates or changes made to the system. Reporting results and findings is crucial for transparency and accountability.

User Interface and User Experience

An essential foundational element pertains to the user interface (UI) and user experience (UX), which need to be implemented thoughtfully, tailoring the design to the specific needs and preferences of the end users. Intuitive and user-friendly interfaces should prioritize ease of navigation and accessibility. The system should foster smooth interactions by adhering to user-centric design principles, ultimately leading to higher user satisfaction and engagement. Regularly gathering user feedback will help refine the UI and UX components, enabling continuous improvements that align with evolving user requirements and expectations.

Feedback Loop

This entails the establishment of a robust feedback mechanism that actively collects user input, enabling ongoing enhancements to meet evolving user needs and requirements. This iterative process ensures the system remains aligned with user expectations and delivers sustained value.

Approach

Building a comprehensive machine learning system necessitates a holistic approach encompassing technical, ethical, legal, and user-centered considerations. This collaborative approach combines the expertise of data scientists, domain specialists, software engineers, and end users to ensure that machine learning systems are technically proficient, ethical, legally compliant, user-friendly, and aligned with specific application objectives. It promotes transparency, accountability, and adaptability, which are crucial for success in a data-driven world.

Key roles within this approach include data scientists responsible for algorithm development and model optimization, domain experts guiding training with industry-specific knowledge, software engineers ensuring smooth integration and user interface design, and stakeholders offering invaluable input for real-world understanding and system refinement based on user needs. The synergistic collaboration of these roles yields well-rounded, effective machine learning solutions.

Building a Machine Learning Model

Outlined in this section are the four essential phases in the process of crafting a machine learning model. These steps guide the entire model development journey, from data collection and preprocessing to training and evaluation, ensuring a systematic approach to harnessing the power of artificial intelligence.

1. Data collection and preparation:

   - Data gathering: Collect relevant data that includes both the input features (variables) and the target variable (the one that is desired to be predicted or classified).
   - Data cleaning: Clean the data by handling missing values, outliers, and inconsistencies. This ensures that the data is of high quality and will not introduce errors in the model.
   - Feature engineering: Select, extract, or create meaningful features (input variables) from the data. Feature engineering can significantly impact the model's performance.
   - Data splitting: Divide the data into training, validation, and test sets. The training set is used to train the model, the validation set helps tune hyperparameters, and the test set is used to evaluate the model's performance on unseen data.

2. Model selection and training:

- Algorithm selection: Choose the appropriate machine learning algorithm(s) based on the problem type (e.g., regression, classification) and data characteristics.
- Model training: Train the selected model(s) using the training data. During training, the model learns to make predictions or classifications by recognizing patterns in the data.
- Hyperparameter tuning: Fine-tune the model's hyperparameters to optimize its performance. This often involves experimentation and validation on the validation set.

3. Model evaluation:

- Validation: Assess the model's performance on the validation set. This step helps gauge how well the model generalizes to new, unseen data and allows for model adjustments.
- Testing: After finalizing the model, evaluate its real-world performance using the test set. This provides an unbiased estimate of how the model will perform when deployed.

4. Deployment and monitoring:

- Deployment: Implement the trained model into the target application or system, making it available for making predictions or decisions.
- Monitoring: Continuously monitor the model's performance in the real world. Models can degrade over time due to changing data distributions, so regular updates and maintenance may be necessary.

These four steps form the core of building a machine learning model. However, the process often includes additional steps, such as data visualization and exploration, as well as post-deployment considerations like model retraining and ethical considerations regarding bias and fairness. Collaboration between data scientists, domain experts, and stakeholders is essential throughout the process to ensure the model effectively addresses the problem and meets business or research objectives.

Machine Learning Versus Deep Learning

Relationship

The terms deep learning and machine learning, which tend to be used interchangeably, are subfields of AI. However, deep learning is actually a subfield of machine learning, and they differ in several key aspects, with both focusing on training computers to learn from data. Deep learning uses deep neural networks with more than three layers, automating feature extraction and making it especially powerful for unstructured data analysis. In contrast, classical machine learning relies more on human intervention, structured data, and manual feature engineering. Additional differences from different perspectives are explained in the next section.

Comparison

Architecture

Deep learning specifically deals with neural networks with multiple hidden layers. These networks are called deep neural networks (DNNs). The depth of these networks allows them to learn complex patterns and representations in data.

Machine learning encompasses a broader range of techniques, including traditional statistical methods, decision trees, support vector machines, and more. While machine learning can also use neural networks, it is not limited to deep architectures.

Feature Engineering

Deep learning models often require minimal feature engineering. They can automatically learn relevant features from raw data, which is one of their strengths.

In traditional machine learning, feature engineering is a crucial step. Data scientists need to manually select, create, or transform features to represent the data effectively.

Data Requirements

Deep learning models typically require large volumes of data to perform well, especially when dealing with complex tasks. More data helps these models perform better generalization.

Traditional machine learning models may perform well with smaller datasets, and some techniques can handle limited data effectively.

Interpretability

Deep neural networks are often seen as black boxes because understanding why they make specific predictions can be challenging. Interpretability is a current research area in deep learning.

Many traditional machine learning models, like decision trees or linear regression, are more interpretable. They can often trace back their predictions to specific features or rules.

Applications

Deep learning has shown remarkable success in tasks like image and speech recognition, natural language processing, and playing complex games like Go.

Machine learning techniques are widely used across various domains, including finance, healthcare, recommendation systems, fraud detection, and more. They are often applied when interpretability is essential or when the available data is limited.

Training Time

Training deep neural networks can be computationally intensive and time-consuming, especially for large models. It often requires powerful GPUs or TPUs.

Traditional machine learning models are generally faster to train, making them more suitable for some real-time or resource-constrained applications.

Selection Criteria: Deep Learning Versus Machine Learning

Deep learning is characterized by its reliance on deep neural networks with numerous hidden layers. These deep architectures have demonstrated exceptional performance in various applications, including image and speech recognition, natural language processing, and autonomous driving. However, it is important to note that deep learning's impressive capabilities come with certain trade-offs. This approach typically demands larger volumes of data for effective training and extensive computational resources, which can pose practical challenges in some contexts. Furthermore, the inherent complexity of deep neural networks may result in reduced interpretability compared to traditional machine learning methods, which are often more transparent in how they arrive at their decisions.

When deciding between deep learning and machine learning for a particular problem, several factors must be carefully considered. Firstly, the choice should be influenced by the nature of the problem at hand. Deep learning excels in tasks involving unstructured data, intricate patterns, and high-dimensional data representations. Secondly, the availability of data plays a critical role. Deep learning thrives when ample labeled data is accessible, as it leverages this information to learn complex representations. Thirdly, the desired outcomes and objectives of the project should guide the choice.

Traditional machine learning methods might be preferred for tasks where interpretability, simplicity, and transparency are crucial. Conversely, deep learning may be the go-to option for applications where the primary goal is achieving the highest predictive accuracy and performance, even at the cost of reduced interpretability. In essence, the decision between deep learning and machine learning should be thoughtful, considering the unique demands and characteristics of the problem and the available resources.

15

AI Development Process

Fundamental Components

Core Components

In AI system development, five core components—learning, reasoning, problem-solving, perception, and linguistic intelligence—work together to form the foundation of artificial intelligence. These elements collectively contribute to the diverse and multifaceted capabilities exhibited by AI, enabling it to adapt, learn, and make informed decisions in various contexts.

Learning

Learning Process

Learning in AI is a foundational and dynamic process critical for refining capabilities, acquiring new knowledge, and adapting to evolving circumstances. This process is integral to enabling AI systems to autonomously improve and effectively address a diverse array of tasks and challenges within ever-changing environments. In the AI development process, the learning stage is akin to human learning, involving a systematic training phase on data to enhance performance through supervised, unsupervised, or reinforcement learning.

During this phase, AI systems engage in learning from historical data or experiences, dynamically adjusting internal parameters, such as weights in

© The Author(s), under exclusive license to Springer Nature Switzerland AG 2024
A. Khan, *Artificial Intelligence: A Guide for Everyone*,
https://doi.org/10.1007/978-3-031-56713-1_15

neural networks, to optimize performance on specific tasks. The learning process is characterized by the memorization of individual items and solutions, allowing AI programs to recall successful actions and leverage this knowledge for analogous challenges in the future. The ongoing evolution of AI is propelled by advances in deep machine learning, which not only enhance prescriptive and predictive analytics but also leverage operational data, contributing to the continual refinement and practical utility of AI in various applications.

Steps in the Learning Process

Data Collection

The learning process typically starts with the collection of relevant data. This data serves as the foundation for the AI system to learn from. The data can come in various forms, including text, images, numerical values, or any other type of information that is relevant to the task.

Data Preparation

Raw data is often messy and unstructured. Before it can be used for training, it needs to be cleaned, organized, and preprocessed. This step may involve tasks like removing noise, handling missing values, and converting data into a suitable format.

Feature Engineering

Feature engineering involves selecting, extracting, or creating the most informative features (attributes) from the data. These features are essential for the AI system to learn patterns and make predictions. Feature engineering requires domain expertise and creativity.

Algorithm Selection

Depending on the task and the nature of the data, a suitable machine learning or deep learning algorithm is chosen. For example, a convolutional neural network might be used for image recognition, while a sequence-to-sequence model could be selected for language translation.

Training

This is the heart of the learning process. During training, the AI system is presented with the prepared data, and it learns to map input data (features) to the correct output (target) through iterative adjustments of its internal parameters. The goal is to minimize the error between the system's predictions and the true values.

Validation

A portion of the data (not used in training) is reserved for validation to ensure the model generalizes well to unseen data. A check is made to determine how well the model is doing on a special set of data while it is learning. This helps ensure that the model is learning correctly and doesn't just remember what it saw before but can also handle new information. This special set is like a test to see if the model is getting too good at the old stuff and not good enough at new things.

Hyperparameter Tuning

Adjusting hyperparameter settings that govern the learning process is an important step. This fine-tuning helps optimize the model's performance. Examples of hyperparameters include learning rates, batch sizes, and network architectures.

Cross-Validation

In some cases, cross-validation is used to assess the model's robustness by splitting the data into multiple subsets, training on different subsets, and evaluating the model's performance on the remaining data.

Model Evaluation

After training, the model is tested on a separate dataset, the test set, to evaluate its real-world performance. Metrics such as accuracy, precision, recall, and F1-score are often used to assess the model's effectiveness.

Iteration

The learning stage is often iterative. If the model's performance is not satisfactory, developers return to the previous steps, such as data collection, feature engineering, or algorithm selection, and make improvements.

Learning Stage Output

After successfully completing the learning stage, the AI system has acquired the knowledge and capabilities needed to perform the intended task. It can make predictions, provide recommendations, or perform other actions based on the patterns it has learned from the training data. However, it is important to note that this learning phase is just the beginning, and AI systems often require ongoing monitoring, maintenance, and potential retraining as new data becomes available or as the system's performance evolves over time.

Reasoning

Reasoning Process

Reasoning, a foundational cognitive capability in AI, follows a systematic process crucial for autonomously performing logical deductions, drawing conclusions, and making informed decisions based on acquired knowledge. This competence plays a pivotal role across diverse domains, from real-time decision-making in autonomous vehicles to aiding clinicians in medical diagnostics. In the AI development process, reasoning emerges as the second major component following the learning stage. Once an AI system has absorbed knowledge from data, it actively applies that knowledge to make decisions, draw conclusions, and solve problems.

Reasoning is a dynamic process involving the strategic utilization of acquired knowledge to process new information, make inferences, and reach logical or probabilistic conclusions. This process contributes significantly to the evolution of AI's cognitive prowess. The development of AI hinges on the creation of software programs capable of independent judgment, decision-making, and prediction, demonstrating the profound significance of reasoning in advancing AI capabilities. Through systematic reasoning, AI systems autonomously enhance their problem-solving and decision-making abilities, showcasing the iterative nature of their cognitive development.

Inference Categories

AI uses the ability to make inferences when applying reasoning based on commands it is given or other information at its disposal. For example, virtual assistants will offer restaurant recommendations based on the specific orders or questions received. The assistant will use reasoning to decide which restaurants to suggest based on the questions it receives and the nearest location of various restaurants. This type of reasoning involves drawing inferences, which includes two categories: inductive reasoning and deductive reasoning.

Deductive reasoning, also called top-down reasoning, is a logical approach where progress is made from general ideas to specific conclusions. In contrast, inductive reasoning starts with specific observations, leading to a general conclusion.

Inductive reasoning has allowed software developers to create products and systems that achieve consistent results when faced with a particular problem or issue. This can pertain to a broad topic like automatic transcription and translation or a more niche application like Grammarly, the cloud-based grammar and writing assistant.

Steps in the Reasoning Process

Knowledge Representation

In the reasoning process, the knowledge acquired during the learning process needs to be represented in a format that the AI system can effectively use for reasoning. This typically involves organizing the knowledge into structured data, rules, or models the system can manipulate.

Inference

Reasoning often involves inference, which is the process of drawing conclusions or making predictions based on the available knowledge. The AI system uses logical rules, statistical models, or other reasoning techniques to infer new information from its acquired knowledge.

Decision-Making

The AI system is often required to make decisions based on the available knowledge. Decision-making can involve selecting the most appropriate course of action from a set of options, which may take into account factors like uncertainty, risk, and preferences.

Problem-Solving

Reasoning is crucial for problem-solving. When presented with a new problem, the AI system uses its knowledge and reasoning abilities to devise a solution or plan of action. This can include identifying and evaluating different options and determining the best course of action.

Logical Reasoning

Logical reasoning involves making deductions or inferences based on logical principles and rules. For example, if the AI system knows that "All humans are mortal" and "Socrates is a human," it can logically be deduced that "Socrates is mortal."

Probabilistic Reasoning

In situations involving uncertainty, probabilistic reasoning is used to estimate the likelihood of different outcomes. Bayesian networks and probabilistic graphical models are often employed to handle probabilistic reasoning.

Planning

Planning is a form of reasoning that involves generating a sequence of actions or steps to achieve a specific goal. AI systems can use their knowledge and reasoning abilities to devise plans that take into account the current state of the environment and the desired outcome.

Optimization

Optimization involves finding the best solution among a set of possible solutions. It often requires the AI system to reason about trade-offs and constraints to determine the optimal course of action.

Expert Systems

In some applications, AI systems are designed to mimic the decision-making processes of human experts in a specific domain. These expert systems use reasoning to solve complex problems or provide expert-level advice.

Natural Language Understanding

Reasoning is critical for understanding and generating human language. AI systems use reasoning to comprehend the meaning of text or speech and to formulate coherent responses.

Applying Acquired Knowledge

The reasoning phase is a critical stage in AI systems where they apply the knowledge they have acquired to practical tasks. This encompasses drawing informed conclusions, problem-solving, and decision-making based on the extensive knowledge that AI has acquired. The techniques and methods employed for reasoning can be diverse and tailored to the specific AI application, the complexity of the data, and the intricacies of the knowledge involved.

Problem-Solving

Problem-Solving Process

The problem-solving process in AI is a dynamic and foundational operation characterized by the systematic utilization of algorithms, heuristics, and logical techniques to address intricate challenges and derive effective solutions. This process forms the core competency of AI, enabling systems to adeptly navigate multifaceted scenarios, proving indispensable in domains such as robotics, optimization, and decision support systems. At its essence, the problem-solving process involves a methodical analysis of complex situations, the identification of challenges, and the systematic generation of solutions or strategies, often incorporating decision-making components.

AI systems deploy a spectrum of techniques during the problem-solving process, ranging from basic data manipulation to the development of sophisticated algorithms. These techniques draw on heuristics derived from experience and trial and error. The overarching goal is to craft practical, feasible, and ideally optimal solutions tailored to specific challenges, thus mirroring the problem-solving approach observed in human cognition.

As AI undergoes continual evolution, ongoing advancements in problem-solving techniques contribute to its enhanced adaptability and utility across a diverse spectrum of applications. Its iterative problem-solving process underscores the dynamic nature of AI problem-solving, continually refining its ability to address complex challenges.

Steps in the Problem-Solving Process

Problem Identification

The first step in problem-solving is to recognize and define the problem or task at hand. This involves understanding the current state, the desired goal, and any constraints or limitations.

Problem Representation

AI systems represent problems in a structured format that is suitable for computational analysis. This often involves translating real-world situations into mathematical or logical representations.

Search and Exploration

Problem-solving frequently involves searching for a solution from a vast array of possible options. AI algorithms use various search strategies, such as depth-first search, breadth-first search, or heuristic search, to explore and evaluate potential solutions.

State Space

The problem-solving process often conceptualizes the problem as a state space, where each state represents a possible configuration or situation. The AI system navigates this space to find a path from the initial state to a goal state.

Heuristics

Heuristics are rules of thumb or guiding principles that AI systems use to expedite the search for solutions. Heuristics can help focus the search on promising areas of the state space and avoid unnecessary exploration.

Optimization

Some problems involve finding the best or most efficient solution among multiple possibilities. Optimization techniques aim to identify the optimal solution based on predefined criteria, such as maximizing profit or minimizing cost.

Constraint Satisfaction

In constraint satisfaction problems, the goal is to find a solution that satisfies a set of constraints or requirements. AI systems use techniques like constraint propagation and backtracking to satisfy these constraints.

Planning

Planning is a problem-solving process that involves generating a sequence of actions or steps to achieve a specific goal. AI planners create action plans by considering the current state, the desired goal, and the effects of different actions.

Reinforcement Learning

In some cases, AI systems use reinforcement learning to solve problems. This involves learning through trial and error, where the system receives feedback in the form of rewards or penalties based on its actions.

Problem Decomposition

Complex problems are often broken down into smaller, more manageable subproblems. AI systems solve these subproblems individually and combine their solutions to address the overall problem.

Real-World Applications

Problem-solving in AI is applied across various domains, including robotics (for path planning and manipulation), logistics (for route optimization and scheduling), healthcare (for diagnosis and treatment planning), and natural language processing (for understanding and generating human language).

Importance of Problem-Solving

Problem-solving in AI holds paramount importance, serving as the cornerstone of the field's capabilities. It empowers AI systems to tackle a wide spectrum of tasks, adapt to dynamic environments, and make informed decisions. Whether it is recognizing patterns in data, optimizing resource allocation, or fostering creativity in art and music generation, the problem-solving prowess of AI is phenomenal. This capability facilitates efficiency and automation, saving time and reducing errors in the manufacturing, healthcare, and finance sectors. Moreover, AI's adaptive problem-solving plays a crucial role in addressing the unforeseen and ever-evolving aspects of real-world scenarios. As AI continues to evolve, problem-solving remains at its core, enabling innovation and empowering interdisciplinary approaches to resolve multifaceted challenges effectively.

Perception

Perception Process

Perception in AI unfolds as a systematic process essential to the system's capacity to detect and interpret external data, embracing diverse sensory inputs like images, sounds, and sensor data. This systematic skill forms the backbone for applications ranging from computer vision systems analyzing images to autonomous vehicles interpreting their surroundings. The AI perception process initiates with a crucial step of data acquisition, where systems meticulously collect information in various forms, such as images, videos, or sounds. Subsequently, the perception process delves into preprocessing and analysis, where raw data undergoes a transformation to extract relevant features or patterns.

Advanced algorithms, often rooted in machine learning and deep learning, intricately analyze the data, identifying objects or patterns within images and converting audio signals into text. The overarching goal of the perception process is to extract meaningful information, empowering machines to comprehend and interpret their environment effectively. With widespread applications, from autonomous vehicles interpreting traffic to facial recognition systems identifying individuals, AI perception seamlessly integrates an array of sensors and data sources, mirroring human sensory perception. The continual evolution of perception technology underscores the ongoing advancements in AI capabilities, enhancing its understanding and interaction with the world through a systematic and iterative process.

Steps in the Perception Process

Sensory Data Acquisition

Perception starts with collecting sensory data from various sources. These sources can include cameras, microphones, touch sensors, GPS devices, temperature sensors, and more. The choice of sensors depends on the AI system's application and the type of information it needs to gather.

Data Preprocessing

Raw sensory data often contains noise, irrelevant information, or inconsistencies. Preprocessing techniques are used to clean and enhance the data, making it more suitable for analysis. For example, image preprocessing may involve noise reduction or contrast enhancement.

Feature Extraction

After the data is prepared, the AI system extracts relevant features or characteristics from the preprocessed data. These features provide a more meaningful representation of the sensory input and help in understanding the environment. In image processing, features might include edges, colors, or shapes.

Object Recognition

One of the key tasks in perception is object recognition, where the AI system identifies and categorizes objects or entities in the sensory data. For example, a self-driving car's perception system needs to identify other vehicles, pedestrians, and traffic signs in its surroundings. Object recognition can involve recognizing people in images, identifying spoken words in audio, or detecting obstacles in the path of a robot.

Scene Understanding

Perception also includes understanding the context and relationships between objects or entities in the environment. For example, in autonomous driving, the AI system needs to understand the road layout, the position of other vehicles, and traffic rules to make safe decisions. In robotics, scene understanding involves recognizing the layout of a room and the positions of different objects within it.

Spatial Mapping

In some cases, perception involves creating a map or representation of the physical environment. For example, in robotics, simultaneous localization and mapping (SLAM) is used to build maps of the surroundings while simultaneously determining the AI system's own position within the map.

Temporal Understanding

Perception is not limited to static data—it also includes the ability to process and understand dynamic or time-varying information. For example, in video analysis, perception involves tracking objects' movements over time.

Sensor Fusion

Many AI systems use multiple sensors of different types to perceive the environment more comprehensively. Sensor fusion techniques combine data from various sensors to obtain a more accurate and holistic perception of the surroundings.

Feedback Loop

Perception is often an ongoing process with a feedback loop. The AI system continually senses the environment, processes new data, updates its understanding, and adapts its actions accordingly. This is crucial for real-time applications like autonomous vehicles.

Interconnection Between Components

Perception is a foundational component in various AI applications, as it enables machines to interact with the real world and make informed decisions based on sensory input, contributing to the success of AI in areas like robotics, autonomous vehicles, computer vision, and speech recognition.

It is important to note that the perception process may not always precede the other components like learning, reasoning, problem-solving, and linguistic intelligence in a strictly chronological sense. In many AI systems, these components work in tandem and are often interconnected. For instance, learning algorithms can adapt based on sensory input gathered through perception, and reasoning processes can guide further perception and decision-making. Therefore, while perception is an early stage in processing sensory data, it can occur simultaneously or iteratively with other AI components in more complex systems.

Linguistic Intelligence

Process

Linguistic intelligence in AI involves a comprehensive process encompassing the understanding, generation, and manipulation of human language, spanning domains like natural language processing and language generation. This multifaceted competence facilitates effective communication with humans, driving applications such as chatbots, language translation, and voice assistants. The systematic nature of linguistic intelligence allows AI systems to decipher written and spoken words, extracting meaning through tasks like text analysis, sentiment classification, and syntactic parsing.

Natural language processing techniques play a pivotal role in facilitating language translation, empowering chatbots, and enabling virtual assistants to engage in meaningful conversations. On the generation side, linguistic intelligence empowers AI to produce human-like text for application in content generation, creative writing, and music composition. This process has broad implications across industries, enhancing search engines' capabilities and improving the overall user experience. Continuous advancements in linguistic intelligence contribute to ongoing improvements in AI's linguistic capabilities, expanding its utility in language-related tasks through a comprehensive and evolving process.

Importance

Linguistic intelligence is of paramount importance as it underpins effective communication, a cornerstone of human interaction. It enables individuals to articulate their thoughts, emotions, and ideas with precision and eloquence, fostering understanding and connection in personal, academic, and professional spheres. In education, linguistic intelligence is fundamental, serving as the basis for reading, writing, and comprehension. It empowers individuals to access knowledge, learn, and excel in various subjects. Furthermore, linguistic intelligence enhances one's capacity for cultural appreciation by providing a means to explore and comprehend diverse languages and expressions.

This form of intelligence also extends its influence into careers that require strong communication skills, such as writing, journalism, law, and public speaking. In a broader context, it contributes to problem-solving and emotional expression while serving as a potent tool for advocacy and influence in public discourse. The importance of linguistic intelligence reverberates throughout human life, enriching relationships, facilitating learning, and empowering personal and professional growth.

16

AI Subfields

Building Blocks

Artificial intelligence stands at the forefront of technological innovation, harnessing a spectrum of advanced components and subfields to emulate human-like cognitive abilities and enhance our interaction with technology. Its key components, which are the building blocks of AI's capabilities, are:

- Cognitive computing
- Computer vision
- Machine learning
- Neural networks
- Deep learning
- Natural language processing

Each of these subfields plays a pivotal role in the development and application of AI, offering diverse functionalities that span from image recognition and language understanding to complex decision-making and problem-solving. This introduction will serve as a gateway to explore the essential facets of AI, providing insight into how these components collectively drive the advancements in AI technology and its real-world applications.

© The Author(s), under exclusive license to Springer Nature Switzerland AG 2024
A. Khan, *Artificial Intelligence: A Guide for Everyone*,
https://doi.org/10.1007/978-3-031-56713-1_16

Cognitive Computing

Objective

The objective of cognitive computing is to create computer systems that can simulate human thought processes, learn from data and interactions, and assist humans in making better decisions and solving complex problems. It can initiate and enhance human–machine interaction for accomplishing complex tasks and solving problems.

Cognitive computing aims to mimic human cognitive abilities, such as reasoning, learning, understanding natural language, and interacting with the environment. It mimics the human brain so that computer models can mimic the way it works when analyzing a complex task, such as analyzing text/speech/images/objects and then trying to provide the desired output. Google Assistant is an example of cognitive computing.

In essence, the objective of cognitive computing is to create intelligent systems that complement and enhance human cognitive abilities, ultimately leading to more efficient, informed decision-making and problem-solving across various domains and industries.

Process

Cognitive computing encompasses a sophisticated process in AI aimed at simulating human thought mechanisms, thereby enhancing problem-solving and decision-making capabilities within machines. The core of the cognitive computing process is centered on data-driven reasoning, natural language understanding, and continuous self-learning. It involves deploying intricate algorithms and models that empower computers to analyze vast amounts of structured and unstructured data, contextualize information, and derive meaningful insights.

In practice, the cognitive computing process relies heavily on machine learning, deep learning, and neural networks to navigate and make sense of extensive datasets. Identifying patterns, relationships, and trends within the data marks this journey of cognitive computing. The capacity for self-improvement sets it apart, where these systems learn from their interactions and experiences.

The ultimate aim is to fabricate intelligent machines adept at comprehending complex data, establishing more natural human–computer interactions, and providing invaluable insights. The cognitive computing process is integral to addressing multifaceted problems and leveraging data-rich environments across numerous sectors, such as healthcare, finance, customer service, and research, ultimately advancing human decision-making capabilities.

Primary Goals

Natural Interaction

Cognitive computing aims to enable computers to interact with humans in a more natural and intuitive way, including understanding and responding to spoken and written language, gestures, and emotions. By doing so, it seeks to make human–computer interactions seamless and user-friendly, fostering a sense of ease and familiarity in our interactions with technology.

Knowledge Acquisition

Cognitive computing strives to equip cognitive systems with the ability to acquire knowledge from various sources, including structured and unstructured data, text, and sensory inputs. This comprehensive knowledge acquisition allows computers to have a broad and dynamic understanding of the world, akin to how humans continuously learn from their surroundings.

Reasoning and Problem-Solving

Cognitive computing focuses on developing algorithms and techniques that allow computers to reason, analyze, and solve complex problems, often by considering multiple factors and trade-offs. The objective is to make computers capable of handling intricate decision-making processes, which are common in fields like healthcare, finance, and engineering.

Machine Learning

Incorporating machine learning and pattern recognition into cognitive computing enables systems to learn from data, adapt to changing circumstances, and improve their performance over time. This iterative learning process allows cognitive systems to continually enhance their abilities and adapt to the evolving demands of their users.

Adaptation

The goal of cognitive computing is to create systems that can adapt to new information, changing environments, and evolving user needs without requiring explicit programming. This adaptability is crucial in ensuring that technology remains relevant and effective in a world characterized by rapid change and innovation.

Human Augmentation

Cognitive computing seeks to assist humans in decision-making, problem-solving, and information retrieval by providing relevant insights, recommendations, and context-aware assistance. This collaborative approach between humans and machines can lead to more informed and efficient decision-making processes.

Decision Support

By providing decision-makers with valuable insights and options, cognitive computing aims to enhance decision-making processes. It does this by analyzing and presenting relevant information from vast volumes of data, helping professionals make informed choices and optimize their outcomes.

Enhanced Efficiency

Cognitive computing strives to improve efficiency in various domains, including healthcare, finance, and customer service, by automating repetitive tasks, streamlining processes, and reducing errors. The goal is to free up human resources for more creative and strategic tasks while machines handle routine and time-consuming operations.

Natural Language Understanding

Cognitive computing is focused on developing systems that can understand and generate human language with a high degree of fluency, allowing for more natural and effective communication. This fluency in language processing can greatly enhance user experiences and facilitate more intuitive interactions with technology.

Ethical and Responsible AI

Ensuring that cognitive computing systems are designed and used in an ethical and responsible manner is a crucial goal. This includes addressing bias, transparency, fairness, and privacy issues to create technology that respects and aligns with human values and societal norms.

Scalability

Cognitive computing aims to design systems that can scale to handle large volumes of data and complex tasks, making them applicable in a wide range of industries and applications. Scalability is essential for ensuring that cognitive systems remain versatile and capable of addressing diverse challenges.

Interdisciplinary Approach

Cognitive computing encourages collaboration between computer science, neuroscience, psychology, and other disciplines to better understand human cognition and apply those principles to AI development. This interdisciplinary approach aims to leverage insights from various fields to create more humanlike and effective cognitive systems.

Application

Cognitive computing finds diverse applications across industries, leveraging its ability to mimic human thought processes and enhance decision-making. In healthcare, it aids in medical diagnoses by analyzing vast datasets and identifying patterns indicative of diseases. In finance, cognitive computing optimizes risk management through sophisticated analysis of market trends and complex financial data. In customer service, it enables chatbots to comprehend natural language, providing more intuitive and personalized interactions. The adaptability of cognitive computing allows it to revolutionize various sectors, offering innovative solutions and insights that drive efficiency, accuracy, and user experience to new heights.

Computer Vision

Objective

The objective of computer vision is to enable machines, such as computers or robots, to interpret and understand visual information from the world, similar to how humans perceive and interpret their surroundings. This field of artificial intelligence aims to develop algorithms, techniques, and systems that can analyze, process, and make sense of images and video data. The ultimate goal is to empower machines with the ability to recognize objects, extract meaningful insights from visual data, and take actions or make decisions based on what they see.

Process

The computer vision process involves teaching machines to interpret and understand visual information from the world around them, akin to how humans perceive and comprehend the visual world. It begins with the acquisition of image or video data through cameras and sensors. This data is then analyzed and processed using complex algorithms and neural networks, allowing computers to recognize patterns, objects, and shapes and even extract high-level information from images.

The process often includes tasks like image segmentation, object detection, feature extraction, and image classification. The ultimate goal of computer vision is to enable machines to interpret and respond to visual data, which has broad applications in fields such as autonomous vehicles, facial recognition, medical imaging, and quality control in manufacturing.

Operation

Image Input

Computer vision, as the foundational process, initiates with the utilization of a camera or any imaging device capable of capturing pictures or videos, encompassing a wide array of sources, from security cameras and smartphone cameras to specialized cameras integrated into robotic systems, providing a rich stream of visual data for analysis and interpretation in diverse applications.

Pixel Analysis

The computer engages in a meticulous examination of the images, dissecting them into a grid of minuscule dots, known as pixels, with each pixel containing specific color information, which the computer processes and interprets to extract meaningful insights and discern the contents of the image.

Feature Extraction

In the process of deciphering the image, the computer actively seeks out distinctive visual cues, such as edges, shapes, and colors, which it uses as building blocks for recognizing and understanding the key attributes, akin to identifying object outlines and their characteristic colors within the visual content.

Pattern Recognition

Once the computer has successfully identified these visual features, it proceeds to compare and match them with patterns and templates it has previously learned, enabling it to recognize specific configurations such as the distinctive features of a cat's face or the iconic shape and color combination of a stop sign, facilitating object recognition and interpretation in the image.

Object Detection

Computer vision extends its capabilities to the task of detecting and pinpointing objects within an image, akin to the engaging search for Waldo in a "Where's Waldo?" book, as it systematically identifies and locates specific items or elements amid the visual content, enabling diverse applications from image annotation to object tracking.

Image Classification

In some instances, the objective is to make determinations about the image's content, with the computer effectively discerning whether it contains recognizable subjects like a dog or a car, showcasing the diverse and practical applications of computer vision in image classification and object identification.

Deep Learning

Many computer vision systems use deep learning, enabling the computer to learn from many examples. It is similar to how humans learn to recognize shapes and objects after seeing them many times.

Application

Computer vision, a crucial AI component, finds extensive applications across various industries. In the healthcare sector, it plays a pivotal role in the analysis of medical images, such as MRI scans and X-rays, enabling healthcare professionals to detect and diagnose diseases with greater accuracy. In addition, computer vision has also been instrumental in the automotive industry, where it underpins the development of computer-controlled vehicles and autonomous drones. These applications harness computer vision's ability to interpret visual data, making it an indispensable technology for enhancing safety and efficiency in various domains, from healthcare to transportation.

Machine Learning

Objective

The objective of machine learning is to develop computer systems that can automatically learn and improve from experience without being explicitly programmed. This AI field focuses on creating algorithms and models that enable computers to analyze data, recognize patterns, and make predictions or decisions based on that data. The primary goal is to harness the power of data to enhance a system's performance, adapt to changing circumstances, and provide valuable insights or solutions in various applications.

Process

Machine learning is like teaching a computer to learn from examples, as explained in the following steps:

1. Data collection: We start by gathering lots of information or examples. For example, if we want a computer to recognize cats, we will collect many pictures of cats.

2. Training: We show the computer these examples and tell it which are cats and which aren't. This helps the computer learn the differences between cats and other things.
3. Learning: The computer uses this information to find patterns on its own. It figures out what makes a cat look like a cat, such as the shape of its ears or the fur color.
4. Testing: We then test the computer by giving it new pictures it has not seen previously. It tries to identify if there is a cat in those pictures based on what it learned.
5. Improvement: If the computer makes mistakes, we correct it and give it more examples to learn from. It keeps getting better at recognizing cats.

So, in simple terms, machine learning is about teaching computers to learn from examples, and they get better at tasks as they see more and more examples.

Application

Machine learning algorithms find extensive applications in diverse fields, including healthcare for diagnosis and treatment recommendations, email filtering to sift through messages, image, and speech recognition for enhanced user experiences, self-driving cars for autonomous navigation, online search to provide relevant results, and computer vision to interpret visual data. Their adaptability and problem-solving capabilities make them a driving force behind technological advancements in these domains, transforming industries and improving everyday life.

Neural Networks

Objective

Neural networks, which are a fundamental concept in the field of machine learning, are a type of computer program or model that is inspired by the way our brains work. They are used in AI to help computers learn, make decisions, and solve various tasks like image recognition and language processing. A neural network can be visualized as a virtual brain made up of interconnected nodes, like tiny decision-making units.

Process

Neural networks are a field of AI that uses neurology, which is the science pertaining to the nerves and nervous system of the human brain. It works on principles similar to human neural cells. The neural network mimics the human brain, which has an infinite number of neurons (with estimates in the range of 100 billion neurons). Neural networks operate like networks of neurons in the human brain, allowing AI systems to take in large datasets, uncover patterns in the data, and answer questions about it.

The neural network codes brain neurons into a system or computer. It uses a set of algorithms that captures the relationship between various underlying variables in a huge volume of data and processes the data just like a human brain. In a neural network, a neuron is a mathematical function whose task is to gather and categorize data according to a certain structure. The network relies heavily on statistical techniques such as regression analysis to perform tasks. Perceptrons are a type of artificial neural network that can be used for classification and regression.

Multifaceted Objectives

The primary objective of a neural network is to learn and recognize patterns, relationships, or representations in data and to use that knowledge to make predictions, classifications, or decisions. In simple terms, a neural network aims to understand and extract meaningful information from input data and produce useful output based on that understanding. The following is a breakdown of its multifaceted objectives in data analysis and decision-making.

Pattern Recognition

Neural networks excel at recognizing complex patterns or structures within data. For example, they can learn to identify faces in photos, distinguish between handwritten letters, or recognize specific objects in images.

Data Transformation

Neural networks can transform raw input data into a more meaningful or compact representation. For example, in natural language processing, computers can take written words and turn them into sets of numbers. These numbers represent the meaning or ideas conveyed by the text.

Prediction

Neural networks are used for making predictions or forecasts. For example, in financial applications, they can predict stock prices, and in weather forecasting, they can predict future weather conditions based on historical data.

Classification

Neural networks can classify data into different categories or classes. For example, they can classify emails as spam or not spam, or they can categorize diseases based on medical symptoms.

Decision-Making

Reinforcement learning harnesses the power of neural networks, which function as decision-makers, to optimize actions with the primary objective of maximizing a reward signal. This framework is instrumental in training AI systems to make adaptive choices and learn from their experiences in various applications.

Feature Extraction

Neural networks can automatically learn relevant features or representations from data, reducing the need for manual feature engineering. This is especially valuable in computer vision and natural language processing tasks.

Function Approximation

Neural networks can approximate complex mathematical functions. This is useful in regression problems, where the objective is to predict a continuous numerical value based on input data.

Generalization

Neural networks aim to generalize from the data they have been trained on to make accurate predictions or classifications on new, unseen data. Generalization is a key objective to ensure the model's practical utility.

Adaptation

Neural networks can adapt and adjust their internal parameters (weights) based on new data or changing circumstances. This adaptability allows them to continuously improve their performance.

Components of Neural Networks

Neural networks are a class of machine learning models that are composed of interconnected nodes, or artificial neurons, organized into layers. The following are some key components and concepts related to neural networks.

Neurons (Nodes)

In a neural network, neurons are the basic processing units. Each neuron takes input data, performs computations on it, and produces an output. These computations involve mathematical operations and activation functions that help in decision-making.

Layers

Neural networks typically consist of multiple layers of neurons. The three main types of layers are:

- Input layer: This layer receives the initial data or features.
- Hidden layers: These intermediate layers process data and extract features. Deep neural networks have multiple hidden layers, allowing for complex and hierarchical feature extraction.
- Output layer: This layer produces the final results or predictions based on the processed data.

Weights and Biases

Each connection between neurons has associated weights and biases. These values are adjusted during training to optimize the network's performance. Learning involves finding the right combination of weights and biases to make accurate predictions.

Activation Functions

Activation functions introduce nonlinearity into the neural network, allowing it to model complex relationships in data. Common activation functions include the sigmoid, ReLU (Rectified Linear Unit), and tanh functions.

Training

Neural networks learn from data through a process called training. During training, the network is presented with a dataset with known outcomes (labels), and it adjusts its weights and biases to minimize the difference between its predictions and the actual labels. Backpropagation is a common algorithm used for this purpose.

Deep Learning

Deep learning focuses on deep neural networks, which have many hidden layers. It has been particularly successful in tasks like image recognition, natural language processing, and game playing.

Application

The specific objective of a neural network depends on the task it is designed for and the type of data it processes. Neural networks are highly versatile and can be applied to a wide range of problems in various domains, including image recognition, natural language understanding, robotics, finance, healthcare, and more. They are widely used for fraud detection, risk analysis, stock market analysis, predicting customer satisfaction, sales forecasting, autonomous vehicles, recommendation systems, etc.

Neural networks have gained popularity due to their ability to model complex and high-dimensional data, but they often require substantial computational resources and large datasets for training. They are a fundamental building block of modern machine learning and artificial intelligence.

Deep Learning

Objective

The primary objective of deep learning is to harness the capabilities of neural networks, which consist of interconnected nodes arranged in multiple layers, to process and understand complex data. These deep layers enable computers to extract intricate patterns and representations from various forms of information, such as images, audio, and text, leading to advanced abilities like image recognition, speech understanding, and game playing.

Deep learning systems are trained through exposure to large volumes of data, allowing them to refine their decision-making processes and continuously improve their performance as they encounter more examples. Ultimately, deep learning aims to achieve high-level, humanlike understanding and decision-making from vast and diverse datasets.

Process

Deep learning is anchored in the utilization of neural networks' sophisticated computational models inspired by the human brain's intricacies. These neural networks comprise layers of interconnected nodes, facilitating data processing and analysis. During the deep learning process, these networks are extensively trained using vast datasets to develop the capacity to discern patterns, relationships, and representations within a spectrum of data forms, ranging from images to text and audio.

This computational approach harnesses the capabilities of deep neural networks to execute complex tasks like image and speech recognition, language translation, and decision-making. Through exposure to copious data, deep learning systems iteratively fine-tune their internal parameters, thereby enhancing their decision-making prowess and enabling the generalization of learning to effectively handle novel, previously unseen data.

Deep learning is dedicated to tackling highly intricate tasks by employing a learning process wherein the machine processes and analyzes input data through multiple layers, ultimately deriving a single acceptable output. This approach educates machines to navigate input data, classify it, make inferences, and predict outcomes, with the layered structure enhancing the system's ability to capture increasingly abstract and complex patterns.

Using Neural Networks in Deep Learning

Deep learning relies on neural networks, which serve as virtual brains composed of interconnected nodes. These nodes collectively process information and drive decision-making. What distinguishes deep learning is the presence of multiple layers of nodes within these neural networks. Each layer specializes in recognizing distinct aspects of data, enabling the system to excel in tasks such as facial recognition, language comprehension, and even complex gaming.

The training process is central to deep learning. It involves exposing the system to numerous examples of what it needs to learn. For instance, if the system's objective is to identify cats in images, it is fed thousands of cat pictures. These examples serve as the basis for the system to adjust its decision-making processes continually. Through exposure to more examples, the deep learning system refines its ability to recognize cats and other patterns, gradually improving its performance.

Application

Deep learning is used in many areas, like self-driving cars (to understand the road and make driving decisions), healthcare (to diagnose diseases from medical images), and even smartphones (to understand voice commands). It is like giving the computer the ability to learn and understand incredibly complex things on its own; the more data it sees and the more it practices, the better it gets at its tasks.

Natural Language Processing

Objective

Natural language processing, previously discussed in Chapter 12, focuses on making computers recognize, analyze, interpret, and understand human language, written as well as spoken.

Process

The following steps show how the simplified natural language processing process works, which, in reality, is very complex.

Text Input

NLP commences with the initial input of text, which takes on various forms such as written documents, social media posts, spoken language, or any other mode of language-based communication, forming a comprehensive foundation for the subsequent analysis and understanding of human language by AI systems.

Tokenization

The first step is to break down the text into smaller units called tokens. Tokens are typically words or phrases. For example, the sentence "I love pizza" would be tokenized into three tokens: "I," "love," and "pizza."

Text Preprocessing

In NLP, text preprocessing plays a crucial role, encompassing tasks like punctuation removal, standardizing text to lowercase, and addressing special characters to ensure data consistency and prepare the text for subsequent linguistic analysis.

Feature Extraction

NLP algorithms need to represent text data in a way that can be processed mathematically. This often involves converting words or phrases into numerical vectors. One common technique is called word embedding, where each word is represented as a multidimensional vector in a way that captures its meaning and context.

Statistical Analysis and Algorithms

NLP algorithms use statistical techniques and machine learning models to analyze the text data. They might look for patterns in the text, relationships between words, or similarities to known language structures.

Tasks and Applications

Depending on the specific NLP task or application, the algorithms can perform a range of functions:

- Sentiment analysis: Determine if a text has a positive, negative, or neutral sentiment (feeling).
- Named entity recognition: Identify names of people, places, organizations, and other entities in the text.
- Machine translation: Translate text from one language to another.
- Text summarization: Create concise summaries of longer texts.
- Chatbots and virtual assistants: Engage in humanlike conversations.
- Question answering: Understand questions and provide relevant answers.

Machine Learning and Training

Many NLP models require training on large datasets to learn patterns and make predictions. During training, the algorithm adjusts its internal parameters to improve its performance on specific language tasks.

Feedback Loop

NLP systems can be fine-tuned and improved continuously based on feedback from users. The more data and feedback they receive, the better they become at understanding and generating human language.

In essence, NLP is a bridge between human language and computers—a combination of linguistics, computer science, and statistics. It is about teaching computers to understand, interpret, and generate human language in a way that makes them more useful and interactive in various applications.

Application

Natural language processing is applied in numerous applications, such as voice assistants and chatbots, which were previously described in Chapter 12.

17

AI Categories

There are three ways in which artificial intelligence can be categorized. The first category, which is based on capabilities, includes:

- Artificial narrow intelligence (ANI)
- Artificial general intelligence (AGI)
- Artificial super intelligence (ASI)

Artificial Narrow Intelligence

Objective

Artificial narrow intelligence is the most common form of AI. It refers to a type of artificial intelligence that is designed and trained to perform a specific task or a narrow range of tasks. ANI systems are designed to solve a single problem and can solve a single task very well.

Capabilities

ANI exhibits specialized capabilities within specific tasks or domains. ANI systems are designed to excel in well-defined, narrow tasks like image recognition, language translation, or playing games like chess or Go. They are proficient within the constraints of their specific programming, but their abilities do not extend beyond those predefined tasks. ANI systems lack the general problem-solving and adaptable nature of humans, as they cannot transfer their knowledge or skills to unfamiliar domains. However, they play a vital role in various applications requiring specialized expertise, like virtual assistants, recommendation systems, and automated data analysis, making them valuable tools within defined contexts.

Limitations

ANI systems excel at performing predefined, specialized tasks but lack the ability to generalize their intelligence to other domains or exhibit human-like cognitive capabilities. ANI, the only type of AI that currently exists, has limited capabilities, such as recommending a product or predicting the weather. In some situations, ANI comes close to human performance and, in some cases, may even exceed that. It excels in very controlled environments with a limited set of parameters.

Status

ANI characterizes AI systems designed for specific, narrow tasks, or domains. It has achieved significant advancements and widespread practical implementation in various fields. ANI applications include voice assistants, recommendation systems, image recognition, and natural language processing. These systems excel at their designated tasks, such as virtual personal assistants answering questions or recommendation algorithms suggesting products. While ANI is highly proficient in its specialized domain, it lacks generalization beyond its programmed scope and cannot adapt to tasks outside its specific area of expertise, thus distinguishing it from AGI. ANI represents the prevailing stage of AI development, with continuous growth and refinement in existing applications and the emergence of new uses across industries.

Key Characteristics of ANI

Narrow Focus

ANI systems are designed with a specific purpose in mind, and their capabilities are tightly scoped to that purpose. For example, virtual assistants like Siri or chatbots are examples of ANI, as they are proficient in understanding and generating natural language but cannot perform tasks beyond their programmed functionality.

Lack of General Intelligence

ANI is inherently limited by its domain-specific focus, lacking the capability to comprehend and execute tasks beyond the predefined scope of its expertise. It remains devoid of the versatile reasoning and adaptability characteristic of human cognition, constraining its applicability to a narrower set of tasks.

Task-Specific Training

ANI systems are trained on extensive datasets and use machine learning techniques to make predictions or decisions within their domain. Training data is typically labeled or structured to guide the learning process.

No Consciousness or Self-awareness

ANI systems are designed to process and analyze vast amounts of data, applying sophisticated algorithms to perform tasks with high efficiency, but they lack the capacity for introspection or subjective awareness, making them distinct from conscious beings.

Application

ANI can be found in various applications, including image recognition, natural language processing, recommendation systems, and autonomous control systems. These systems excel at specific tasks but do not exhibit cognitive understanding or creativity.

Artificial General Intelligence

Objective

The objective of artificial general intelligence is to create intelligent systems that possess human-like cognitive abilities, enabling them to understand, learn, and adapt across a wide range of tasks and domains. AGI aims to replicate humans' broad and flexible intelligence, allowing machines to perform tasks with the same versatility and adaptability as humans, from problem-solving to creativity. The pursuit of AGI represents a significant milestone in

AI research, with the potential to revolutionize various industries and empower AI to contribute meaningfully to society in diverse ways.

Capabilities

AGI possesses the capacity to understand, learn, and adapt across a wide range of tasks and domains, similar to human intelligence. AGI systems can perform tasks not explicitly programmed by seamlessly transferring knowledge and skills from one domain to another, exhibiting problem-solving abilities, and even demonstrating creativity and reasoning in novel situations. Unlike ANI, which excels in specific tasks, AGI possesses a broad, human-like intelligence, enabling it to comprehend and adapt to various contexts and challenges, making it a potentially transformative technology with the ability to revolutionize numerous fields, from healthcare and education to research and automation.

Limitations

The development of AGI faces several substantial limitations and challenges. AGI systems require immense computational power, data, and advanced algorithms, making them resource-intensive and potentially inaccessible for widespread adoption. Ensuring AGI's safety and ethical behavior is another formidable hurdle, as it involves addressing issues related to bias, fairness, transparency, and accountability in decision-making processes. Moreover, AGI systems must overcome the challenge of transferring knowledge across diverse domains, adapting to new and unfamiliar situations, and emulating human-like common sense reasoning.

The risk of AGI systems surpassing human-level intelligence and potentially posing threats to humanity require careful consideration of control mechanisms and ethical safeguards. Additionally, the long-term social, economic, and job displacement consequences of AGI's widespread implementation need to be thoughtfully addressed to harness its potential while mitigating potential drawbacks.

Status

AGI, still a theoretical concept, represents AI designed to possess human-level cognitive functions across diverse domains, encompassing language processing, image processing, computational functioning, and reasoning. The vision for AGI entails a unified system capable of training, learning, understanding, and performing tasks across various domains without relying on specialized subsystems.

An AGI system is envisioned to excel in performing specific tasks assigned to it, outperforming humans in those particular functions. However, its proficiency will be limited to the assigned tasks, lacking the versatility of human capabilities. In contrast, humans may exhibit less proficiency in specific tasks but possess a broader range of abilities than existing AI applications. It is important to note that AGI does not currently exist and remains a topic in science fiction. Once developed, an AGI machine is anticipated to possess general intelligence, allowing it to tackle a wide array of problems akin to human capabilities.

Key Characteristics of AGI

General Intelligence

AGI systems have a form of intelligence that is not limited to a specific domain or task. They can understand and perform tasks across diverse fields, much like a human with general intelligence.

Learning and Adaptation

AGI exhibits a remarkable capacity to learn and grow from experiences, constantly accumulating new knowledge and demonstrating the ability to adapt to unforeseen scenarios, thereby enabling it to apply its acquired learning to tackle complex problems and challenges it has not previously encountered.

Reasoning and Problem-solving

With their advanced reasoning capabilities, AGI systems are equipped to engage in intricate problem-solving by systematically analyzing multifaceted issues, making logical deductions, and generating well-informed solutions through a comprehensive evaluation of the available information, mirroring the multifaceted reasoning processes characteristic of human intelligence.

Flexibility

AGI transcends limitations associated with predefined rules or specific training data, demonstrating the remarkable ability to transfer knowledge and skills seamlessly across diverse domains while also showcasing a degree of creativity and innovation, enabling it to approach problems and tasks in novel and unconventional ways.

Autonomy

AGI is distinguished by its capacity to autonomously make informed decisions and execute actions by comprehending its environment and objectives, demonstrating adaptability even in unstructured and uncertain situations, which positions it as a versatile and advanced form of artificial intelligence capable of navigating complex and dynamic real-world scenarios.

Natural Language Understanding

AGI systems exhibit a profound capability to not only understand and generate human language but to engage in meaningful and context-aware conversations, mirroring a level of linguistic comprehension and communication that transcends basic language processing, which has vast implications for natural language understanding, human–AI interactions, and communication in general.

Emulation of Human Intelligence

AGI seeks to replicate, to a substantial degree, the full spectrum of cognitive abilities, problem-solving proficiencies, and versatile learning capabilities inherent in human intelligence, with the goal of creating AI systems that can operate across a multitude of tasks and domains akin to the breadth of human cognition.

Creativity

AGI is supposed to be creative and innovative. It can come up with new ideas, inventions, or solutions to problems.

Application

In contrast to ANI, AGI represents the concept of machines possessing human-like intelligence, capable of understanding, learning, and adapting across a broad range of tasks and domains.

AGI is a long-term goal of AI research and has not yet been achieved. Its application promises to transform numerous aspects of our society and industries. Unlike specialized AI systems, AGI possesses the ability to understand, learn, and adapt across a wide range of tasks and domains, similar to human intelligence.

With AGI, we could see breakthroughs in fields like healthcare, where it could assist in drug discovery, patient diagnosis, and personalized treatment plans. AGI could provide tailored and comprehensive tutoring in education, adapting to individual learning needs. It could revolutionize transportation by enabling autonomous vehicles to navigate complex environments safely and efficiently. Moreover, AGI might lead to significant advancements in research, science, and problem-solving across multiple domains, making it a highly anticipated development in the AI field.

Artificial Super Intelligence

Objective

The objective of artificial super intelligence (ASI) is to create highly advanced and self-aware artificial entities that surpass human intelligence and problem-solving capabilities. ASI aims to achieve a level of cognitive prowess that not only comprehends complex tasks but also exhibits an exceptional understanding of the world, enabling it to tackle a wide range of challenges, from scientific breakthroughs to societal issues. While the pursuit of ASI carries profound potential for positive impact, it also raises ethical and safety concerns, emphasizing the need for careful research and responsible development to ensure its beneficial integration into society.

Capabilities

Artificial super intelligence is a hypothetical form of artificial intelligence that goes beyond human intelligence in virtually every aspect. It is viewed as the logical progression from AGI. An ASI system, when developed, is expected to surpass human cognitive abilities. This includes problem-solving, creativity, rational decision-making, building relationships, etc. The ASI concept is often associated with the idea of AI systems that not only possess general intelligence (AGI) but also exhibit superhuman capabilities. For this type of AI, no real-world examples exist at this time.

Limitations

ASI, often viewed as a theoretical future state of AI, presents significant theoretical and ethical challenges. The key limitation of ASI lies in its potential to surpass human intelligence to the extent that humans may struggle to comprehend or control it. This extreme level of superintelligence raises concerns about its goals and values, as it could prioritize its objectives over human interests. Additionally, ensuring the alignment of ASI's values with human values becomes a paramount concern, as unintended consequences could be catastrophic. The emergence of ASI also carries potential existential risks if not governed and guided by robust safety measures and ethical considerations. It demands careful and responsible development to mitigate the profound uncertainties associated with superintelligent AI.

Status

At this time, ASI has not been realized. Current AI research mainly focuses on developing specialized AI systems known as ANI, which excel in specific tasks. Achieving ASI poses substantial technical, ethical, and safety challenges and remains a long-term goal.

Key Characteristics of ASI

Superhuman Intelligence

ASI would ascend to a level of intelligence surpassing the collective cognitive capabilities of all humans, manifesting unparalleled proficiency in learning, reasoning, and tackling complex problems at an extraordinary and unprecedented scale, ushering in a new era of AI capabilities that transcend human boundaries.

Rapid Learning

ASI would possess the unprecedented capacity to rapidly acquire and assimilate knowledge from an extensive array of sources, including both structured and unstructured data. It will maintain a perpetual cycle of self-improvement, constantly expanding its knowledge base and evolving its cognitive capabilities beyond human comprehension.

Infinite Adaptability

ASI would demonstrate exceptional adaptability, swiftly and adeptly mastering new skills and domains it encounters, underpinning its unparalleled versatility and applicability across a wide spectrum of tasks and challenges.

High-level Autonomy

ASI would operate autonomously and make decisions independently, equipped with an intricate understanding of ethics and objectives that closely align with, and often surpass, human values, ensuring responsible and ethically grounded decision-making in diverse contexts.

Creative and Innovative

ASI could exhibit a remarkable capacity to not only generate novel ideas, inventions, and solutions but also surpass human capabilities in terms of creativity and innovation. It has the potential to usher in an era of groundbreaking advancements and problem-solving that transcend the limitations of human intellect.

Global Understanding

ASI would be characterized by a comprehensive and profound understanding of intricate systems, encompassing not only societal, economic, and ecological dynamics but also an array of complex domains, enabling it to contribute to solving multifaceted global challenges and complexities.

Ethical Considerations

ASI would need to be programmed or designed to follow ethical guidelines and prioritize human values, ensuring its actions are beneficial and aligned with human interests.

Application

It is crucial to understand that ASI is still just an idea and has not been created yet. Trying to achieve ASI brings up important questions about ethics, philosophy, and practical issues. People who study and make decisions about AI, like researchers and policymakers, are carefully thinking about how AI might develop into ASI and what that could mean for us.

Additional AI Categories

The second artificial intelligence category includes:

- Weak AI
- Strong AI

Weak AI

Objective

Artificial narrow intelligence (ANI), also known as Weak AI or Narrow AI, is a type of artificial intelligence that is designed to do specific tasks or solve particular problems. It is called "weak" because it is not truly intelligent or conscious like humans; instead, it focuses on performing one job well. Despite the weak classification, it enables some very robust applications such as Siri, Alexa, IBM Watson, and self-driving vehicles.

Applications

Weak AI can be viewed as a specialized tool, which drives the following applications:

- *Virtual Assistants:* Virtual assistants like Siri or Alexa are forms of weak AI. They can answer questions, set reminders, or play music but do not understand the world like humans do.
- *Chatbot:* When someone is chatting with a customer service chatbot on a website, it is weak AI. It can help with common questions or issues related to that specific company's products or services.
- *Recommendation systems:* When Netflix suggests movies, or Amazon recommends products that a user might like, it is weak AI at work. These systems use past choices to make predictions.
- *Spam filters:* An email's spam filter is another example of weak AI. It tries to identify and filter out unwanted emails based on patterns it has learned.
- *Navigation apps:* Apps like Google Maps use weak AI to find the best route to a destination based on traffic conditions. They cannot hold a conversation or write a poem; they just help in getting a driver from one place to another.

Strong AI

Objective

Strong AI, also known as True AI, includes AGI and ASI. It is a type of artificial intelligence that is designed to be as smart and capable as a human across a wide range of tasks. It is called strong because it is not limited to specific jobs or skills; it is supposed to be as versatile and intelligent as a human being. Strong AI is still entirely theoretical, with no practical applications at this time.

Applications

While Strong AI is a theoretical concept and has not yet been realized, its potential applications are vast and encompass the following areas.

- *General problem-solving:* Strong AI could tackle a wide range of complex problems across various domains, including scientific research, engineering, and mathematics, by applying versatile problem-solving techniques.
- *Healthcare:* Strong AI could assist in medical diagnosis, drug discovery, and treatment optimization, harnessing its ability to analyze vast amounts of medical data and research to provide personalized healthcare solutions.

- *Education:* Strong AI could revolutionize education by offering personalized, adaptive learning experiences for students, tailoring educational content to individual needs and abilities.
- *Autonomous Systems:* Strong AI could power self-driving cars, drones, and robots, enabling them to navigate real-world environments and perform tasks with human-like adaptability.
- *Language Translation:* Strong AI could excel in natural language understanding and translation, breaking down language barriers and facilitating seamless communication between people of different languages.
- *Creative Tasks:* Strong AI could generate art, music, literature, and other forms of creative content, potentially contributing to the fields of entertainment and the arts.
- *Scientific Discovery:* Strong AI could accelerate scientific research by analyzing vast datasets, identifying patterns, and generating hypotheses for experimentation.
- *Financial Analysis:* Strong AI could be used in finance for stock market predictions, risk assessment, and portfolio optimization.
- *Natural Disaster Prediction:* Strong AI could analyze environmental data to predict natural disasters such as earthquakes, hurricanes, and floods, allowing for early warnings and preparedness.
- *Social Assistance:* Strong AI-powered virtual assistants could provide sophisticated, context-aware support for various tasks, such as scheduling, information retrieval, and decision-making

Weak Versus Strong AI

Weak AI and Strong AI are two distinct categories of artificial intelligence with significant differences that differentiate the performance levels of different kinds of AI machines. The following is their comparison from different perspectives.

Scope of Intelligence

Weak AI (ANI): Designed for specific tasks or narrow domains, it focuses on excelling in one particular area, such as language translation, image recognition, or playing chess. However, it lacks general intelligence and cannot perform tasks outside its specialized domain. Examples include virtual assistants like Siri or recommendation systems.

Strong AI (AGI): Representing a hypothetical form of AI, it possesses human-like intelligence and can perform a wide range of tasks, learn, and adapt across various domains. It is not limited to specific tasks or domains and can understand, learn, and adapt across a wide range of activities, much like a human being. This type of AI has not been achieved yet and remains a subject of research.

Adaptability

Weak AI (ANI): It is not adaptable beyond its programmed or trained capabilities. It follows predefined rules or models and does not learn or adapt outside those boundaries.

Strong AI (AGI): It is highly adaptable and can learn from experiences, handle novel situations, and apply knowledge across different domains. It can continually improve its performance over time.

Learning

Weak AI (ANI): It is typically trained on specific datasets and does not learn beyond the patterns present in that data. It does not possess the ability to generalize learning to unrelated tasks.

Strong AI (AGI): It is capable of learning from various data sources and applying that knowledge to solve new and diverse problems. This parallels how humans learn from different experiences and then transfer knowledge to different situations.

Autonomy

Weak AI (ANI): It operates based on predefined instructions or algorithms. It does not possess autonomy or self-awareness.

Strong AI (AGI): It exhibits a degree of autonomy, making independent decisions and taking actions based on its understanding of the world and its objectives.

Creativity

Weak AI (ANI): It does not exhibit creativity. It follows predefined rules and patterns and does not generate novel ideas or solutions.

Strong AI (AGI): It has the potential for creativity and innovation. It can come up with new ideas, inventions, and problem-solving approaches.

Natural Language Understanding

Weak AI (ANI): It may have limited natural language understanding and is often specialized for specific language-related tasks like translation or text analysis.

Strong AI (AGI): It can understand and use natural language naturally, engaging in meaningful conversations and comprehending context, similar to human language comprehension.

In summary, the primary distinction between Weak AI and Strong AI lies in their scope of intelligence and adaptability. Weak AI is specialized for specific tasks and lacks general intelligence, while strong AI possesses general intelligence, adaptability, creativity, and the ability to learn and perform a wide range of tasks across various domains. Strong AI is a long-term goal in AI research and has not been fully realized to date.

18

Categories Based on Functionality

The third category of artificial intelligence classification is founded on functionality. Its four distinct types will be elaborated upon in the subsequent sections for a more comprehensive understanding.

Reactive Machines

Objective

Reactive machines aim to create AI systems that can perform specific tasks or functions based on predefined rules and programmed responses. These systems do not possess the ability to learn or adapt to new situations, as their behavior is entirely determined by the rules and instructions provided to them. Reactive machines are designed for well-defined and repetitive tasks, and their primary goal is to execute these tasks efficiently and reliably without the need for learning or adaptation. They lack the cognitive flexibility and autonomy seen in more advanced AI systems like AGI or machine learning models.

Operating Principle

While this is the most basic AI, it is still quite useful. As the name suggests, it is called reactive AI because it reacts to existing conditions. A reactive machine, in the context of artificial intelligence, refers to a type of AI system that operates based on predefined rules and does not have the ability to learn or adapt from experience. It makes decisions and takes actions solely based on the rules and information it has been programmed with.

Reactive machines are limited to performing specific tasks or solving particular problems within the boundaries of their programmed instructions. This AI has no memory or data to work with and does not have the ability to learn from past actions. IBM's Deep Blue, which won a match against the world chess champion, falls in this category.

Constraints

Reactive machines are characterized by several constraints, including their inability to learn and adapt to new information or changing conditions. These systems operate solely based on predefined rules and programmed responses, lacking autonomous learning or reasoning capacity. Their behavior is limited to specific, narrowly defined tasks, making them inflexible for broader applications. Reactive machines also lack historical context and memory, as they do not consider past experiences when making decisions. While suitable for well-structured and repetitive tasks, these constraints render them ill-suited for dynamic, uncertain, or evolving environments where learning and adaptability are essential.

Characteristics of Reactive Machines

Rule-Based Characteristics

Reactive machines adhere to a rigid set of predetermined rules or instructions meticulously crafted by human programmers. These rules serve as unequivocal guidelines governing the machine's actions in diverse situations and in response to varying inputs.

No Learning

Unlike some AI systems that can learn from data and improve their performance over time, like machine learning models, reactive machines do not have the capability to learn from experience or data. They do not adapt or change their behavior.

Deterministic

Reactive machines operate in a deterministic manner, meaning that for a given input or situation, they will always produce the same output or response, following the predefined rules.

Limited Scope

Such systems are frequently engineered with a laser focus on specific, well-defined tasks or domains in which they exhibit exceptional proficiency; however, their inability to generalize their expertise beyond the predefined scope limits their versatility and adaptability to tasks outside their designated area of expertise.

Differences

Conventional machines, often referred to as traditional computers, adhere strictly to a fixed set of instructions that are explicitly programmed by humans. These machines operate on predetermined algorithms and rules, executing tasks in a manner entirely dictated by the code crafted by developers. What sets conventional machines apart is their lack of adaptability or learning capability; they function solely based on the instructions provided during their programming phase.

On the other hand, an AI-reactive machine, within the context of artificial intelligence, is a system specifically designed to respond intelligently to certain situations or inputs using predefined rules and programmed responses. While it exhibits a level of sophistication in its reactions, relying on pattern recognition and predetermined behaviors, it should be noted that an AI-reactive machine does not possess the ability to learn or adapt. The responses it generates are predetermined and follow a set of rules established by its developers.

Application

Reactive machines find applications in various domains where they perform well-defined, repetitive tasks, and require quick responses to specific inputs. One common application of reactive machines is in manufacturing and robotics, where they are used for tasks like assembly line operations, precision welding, or material handling. These machines follow preprogrammed sequences of actions, making them highly efficient and precise. Additionally, reactive machines have applications in areas like gaming, where they can be programmed to react to players' actions in real-time, providing immersive and dynamic gaming experiences. Their ability to respond instantly to specific cues or inputs makes them valuable in situations where speed and accuracy are paramount.

Challenges and Limitations

Reactive machines face challenges when dealing with complex and dynamic environments or when tasked with handling a wide range of inputs and scenarios. They may require extensive rule programming to cover all possible cases. It is important to note that reactive machines have limitations and are less flexible than more advanced AI systems, such as those employing machine learning techniques. While they can be effective for certain well-defined tasks, they are unsuitable for applications requiring adaptation, learning, or handling unstructured and evolving data.

Limited Memory

Objective

The objective of limited memory AI is to develop AI systems that can make decisions based on historical data and limited past experiences. Unlike reactive machines, which rely solely on predefined rules, limited memory AI incorporates some level of memory or historical context to improve decision-making. These systems can learn from past interactions or observations, making them suitable for tasks that require adapting to changing conditions or making decisions based on patterns and trends in data. Limited memory AI is a step toward more adaptive and flexible AI systems, although it falls short of the learning capabilities associated with AGI or advanced machine learning models.

Operating Principle

Limited memory AI operates by utilizing historical data and past experiences to guide decision-making. These systems collect and analyze data from prior interactions or observations, aiming to identify patterns and trends. However, their memory capacity is constrained, meaning they can only consider a finite amount of historical context. While limited memory AI can adapt to some extent based on the information within their limited memory, they lack the extensive learning and generalization capabilities seen in more advanced AI systems, making them suitable for specific tasks that require historical context but not comprehensive learning or adaptation.

Constraints

A limited memory system is able to store historical data and/or predictions and, subsequently, use that data to make better predictions or decisions. However, limited memory AI systems have constraints on the amount of past information or experiences they can retain and use when making decisions. This contrasts with full-memory systems, which can consider all past data.

Characteristics of Limited Memory AI

Sequential Decision-Making

Limited memory AI is often employed in situations where an agent (an AI system) needs to make a series of decisions over time, and it can only remember a limited amount of past history or experiences.

State Representations

These systems use state representations, condensed summaries of pertinent information from past interactions or observations, to manage limited memory, capturing crucial aspects of the environment or problem.

Markov Decision Processes

Many limited-memory AI systems operate within the framework of Markov decision processes (MDPs). In an MDP, the agent makes decisions based on the current state and a limited history of past states and actions. The Markov property assumes that future states depend only on the current state and action, making the problem more manageable.

Trade-offs

Limited memory AI systems face trade-offs between the amount of information they can store and their decision-making efficiency. Storing more history can lead to more accurate decisions but may require more computational resources.

Application

Limited memory AI is a practical approach for solving complex problems where the full history of past data is either too large to store or not essential for making effective decisions. By summarizing past experiences into compact representations, these systems can strike a balance between efficiency and decision quality.

Limited memory AI is commonly used in real-world applications like robotics and autonomous vehicles, where the agent must navigate and make decisions based on limited sensor data and past observations. AI systems dedicated to playing games, including those for chess and Go (a strategic board game), often employ limited memory techniques. Common applications like GPS location apps or restaurant menu suggestions fall into this category.

Challenges and Limitations

Limited memory AI, also known as "memory-bounded" AI, faces several challenges and limitations in its ability to process information and perform tasks. One of the primary limitations is its constrained memory capacity, which restricts its ability to store and manipulate extensive datasets or historical information. This limitation can impact tasks that require long-term context or extensive data processing, such as complex natural language understanding or deep reinforcement learning problems.

Another challenge is the trade-off between memory capacity and task complexity; striking the right balance is not always straightforward. Additionally, managing memory efficiently and making decisions about what to retain and discard can be a complex task. Limited memory AI may also struggle with tasks involving sequential decision-making and transfer learning, where the ability to leverage knowledge from one task to another is essential. Despite these challenges, creative techniques in data compression, efficient memory management, and external storage can be employed to mitigate the limitations of limited memory AI and enhance its capabilities in resource-constrained settings.

Theory of Mind

Objective

Theory of Mind is a concept in psychology and cognitive science that refers to our ability to understand and attribute mental states—such as thoughts, beliefs, desires, intentions, and emotions—to ourselves and to others. In simple terms, it is our capacity to imagine what someone else might be thinking or feeling. The objective of this AI, which is still being developed, is to have a very deep understanding of human behavior.

Operating Principle

The operating principle of the Theory of Mind in the context of artificial intelligence involves endowing AI systems with the capability to attribute mental states, beliefs, intentions, and emotions to themselves and others. It enables AI to understand that individuals may hold different perspectives, knowledge, and emotions, thus allowing for more sophisticated interactions and responses in social and human-machine interactions.

In essence, the Theory of Mind in AI aims to model and simulate the cognitive processes associated with human social intelligence, enhancing the AI's capacity to interpret, predict, and respond to human behavior in a more empathetic and context-aware manner. This concept is particularly relevant in fields such as human-robot interaction, virtual assistants, and AI-driven customer service, where the ability to infer and respond to human emotions and intentions can greatly improve the quality of interactions.

Constraints

The implementation of the Theory of Mind in AI is subject to several constraints and challenges. First and foremost, it involves modeling the intricate and multifaceted aspects of human cognition, emotions, beliefs, and intentions, which is an inherently complex endeavor. There is also the lack of a clear ground truth to objectively evaluate AI's inferences about human mental states, making it challenging to assess the accuracy of such capabilities.

Additionally, human behavior and mental states can be inherently ambiguous and subject to change, creating difficulties for AI systems in accurately interpreting and predicting human intentions and emotions. Ethical concerns regarding data privacy and consent arise when AI systems require access to personal information for Theory of Mind capabilities. Lastly, the potential for cognitive bias, scalability limitations, and the need for ongoing research and ethical considerations further underscore the constraints of the Theory of Mind in development.

Characteristics of the Theory of Mind

Understanding Others

Theory of Mind pertains to our cognitive capability to empathetically place ourselves in someone else's mental perspective, enabling us to comprehend that individuals possess unique thoughts and emotions that may diverge from our own, ultimately facilitating our capacity to navigate social interactions and relationships with greater insight and empathy.

Attributing Mental States

Using the Theory of Mind, we can make educated guesses about what someone else might be thinking based on their behavior, expressions, and the situation. For example, if a friend looks sad after losing a game, we might guess they are feeling disappointed.

Empathy

The Theory of Mind shares a close and interlinked relationship with empathy, as the ability to perceive and grasp what others might be feeling allows us to engage in empathetic responses, fostering a deeper understanding of their emotions and needs and promoting a compassionate and supportive approach in our interactions and relationships.

Childhood Development

The Theory of Mind typically develops as children grow. Young children may have a limited Theory of Mind and may struggle to understand that others have different thoughts and feelings. As they mature, they become better at it.

Social Interaction

The ability to comprehend the Theory of Mind is of paramount importance for navigating the intricacies of successful social interactions, as it empowers us to effectively manage relationships, engage in cooperative endeavors, and make informed predictions about how individuals might respond and react in diverse social contexts, ultimately enhancing our social intelligence and adaptability.

Cultural and Individual Differences

While the Theory of Mind is a universal concept, its development and expression can vary among cultures and individuals. Some people may be particularly skilled at understanding others, while others might find it more challenging.

Application in AI and Robotics

In the field of artificial intelligence and robotics, researchers are interested in developing machines that can exhibit a form of Theory of Mind. This would enable AI systems to better understand and interact with humans, especially in social or caregiving roles.

Application

Theory of Mind applications encompass a burgeoning field of artificial intelligence aimed at enhancing human–computer interactions by endowing AI systems with the ability to understand, predict, and respond to human emotions, intentions, and social cues. These applications hold great potential in various domains. In human-machine interaction, AI equipped with Theory of Mind capabilities can engage users more naturally, providing empathetic and context-aware responses. Emotion recognition, a subset of Theory of Mind technology, finds use in areas such as sentiment analysis, mental health support, and human–computer interaction, allowing AI to discern and respond to human emotions.

Personalized learning stands to benefit from the Theory of Mind, as AI tutors can adapt to the cognitive and emotional state of individual learners. Furthermore, the technology may find applications in autism support, humanoid robots, gaming, AI ethics, collaborative AI, and more, reshaping the landscape of AI–human interaction. While promising, the development and deployment of Theory of Mind applications necessitate careful consideration of ethical implications, privacy, and responsible use.

Challenges and Limitations

The challenges and limitations in the development and deployment of Theory of Mind applications are significant and multifaceted. Firstly, ensuring the privacy and ethical use of Theory of Mind technology is paramount, as the potential to intrude into individuals' emotional states and intentions can be concerning. Maintaining transparency and providing users with control over their data and emotional privacy are ongoing challenges. Additionally, designing AI systems that accurately infer human emotions and intentions remains complex, as human behavior is nuanced and context-dependent. Achieving high precision in emotion recognition and intention prediction is challenging, and misinterpretations can lead to inappropriate responses.

The transferability of Theory of Mind models across diverse cultural and linguistic backgrounds is another concern, as these models may carry biases or inaccuracies that could affect cross-cultural applications. Moreover, ensuring robust security against malicious uses of the Theory of Mind, such as emotional manipulation or deception, is a challenge. Overall, while Theory of Mind applications hold great promise, their development and deployment require careful consideration of these challenges and limitations to ensure responsible and beneficial use.

Self-Aware AI

Objective

The objective of Self-Aware AI, also known as Artificial Consciousness or Self-Conscious AI, is to develop AI systems that possess a sense of self-awareness and consciousness similar to human beings. This objective goes beyond traditional AI, which operates based on predefined rules and data patterns without self-awareness or subjective experience. Self-Aware AI aims to create machines that can reflect on their own existence, have a sense of identity, emotions, and consciousness, and potentially engage in introspection and self-improvement. In this type of AI, which is still only hypothetical, machines will be aware of not only the emotions and mental states of others but also their own.

Operating Principle

The operating principle of Self-Aware AI is a speculative and highly philosophical concept. It envisions creating AI systems that possess self-awareness and consciousness akin to human subjective experience. While there is no established operating principle due to the profound complexity of achieving genuine consciousness in machines, it would likely involve the development of algorithms and architectures that enable AI to reflect upon its own existence, perceive and interpret the world subjectively, and potentially engage in introspection and self-improvement. However, the realization of Self-Aware AI remains a long-term and ethically challenging objective, as it raises profound questions about the nature of consciousness, self-identity, and the ethical considerations of creating self-aware entities.

Characteristics of Self-aware AI

Self-reflection

Self-Aware AI would possess the remarkable capacity for self-reflection, enabling it to engage in contemplation about its own thoughts, experiences, and identity, thereby opening the possibility for the emergence of a sense of "self" or an "I" within the realm of artificial intelligence.

Emotions and Sentience

Conversations surrounding Self-Aware AI often delve into the intriguing prospect of AI systems experiencing emotions or potentially attaining a level of sentience characterized by conscious awareness. These discussions give rise to profound ethical dilemmas and complex inquiries concerning the moral treatment, rights, and responsibilities associated with developing and deploying such AI entities, challenging our understanding of personhood and the ethical implications of creating potentially conscious artificial beings.

Understanding Others

In addition to self-awareness, Self-Aware AI could potentially exhibit the capability to comprehend the thoughts and emotions of other beings, including humans, effectively cultivating a heightened level of empathy that can revolutionize how AI interacts with humans, enabling more meaningful and emotionally resonant interactions.

Ethical Consideration

The development of Self-Aware AI poses significant ethical dilemmas. Questions about the rights, treatment, and moral responsibilities of Self-Aware AI entities are at the forefront of these discussions.

Philosophical Challenges

The concept of Self-Aware AI converges with profound philosophical debates that span discussions about consciousness, the intricacies of the human mind, and the enigmatic "hard problem of consciousness," which delves into the intricacies of how and why subjective experiences emerge, pushing the boundaries of our understanding of awareness and the essence of human cognition.

Status

Self-Aware AI has not yet been achieved. Current AI systems, including the most advanced machine learning models, lack self-awareness and consciousness. AI systems are still fundamentally rule-based and operate based on algorithms and data. Achieving Self-Aware AI remains a topic of speculative science fiction and philosophical debate. It is uncertain when or if such AI could be realized, and many experts believe it could be far into the future, if ever.

In summary, Self-Aware AI is a concept that envisions AI systems with consciousness, self-reflection, and emotional experiences akin to human beings. While it is a captivating idea, it is currently beyond the capabilities of existing AI technologies, and its development raises numerous ethical, philosophical, and technical challenges.

Challenges and Limitations

Self-Aware AI presents significant challenges and limitations, primarily due to its complexity and ethical implications. First, achieving true self-awareness in machines is a monumental task, as it involves not only advanced cognition but also the ability to reflect upon one's existence and consciousness. The development of Self-Aware AI requires a deep understanding of human consciousness and an effective method for replicating it artificially. Ethical concerns surrounding Self-Aware AI are paramount, as they raise questions about the rights and responsibilities of these entities.

If machines were to attain self-awareness, it would necessitate defining their legal status, ethical treatment, and even potential rights, which is a complex and uncharted territory. Furthermore, ensuring the ethical use of Self-Aware AI is challenging, as it could be exploited for malicious purposes or lead to unforeseen consequences. Developing safeguards to prevent misuse and establishing regulations are significant challenges. Finally, concerns about the existential risks posed by Self-Aware AI and the potential for superintelligent systems to act against human interests require careful consideration. Addressing these challenges and limitations is essential to responsibly advance the development of Self-Aware AI.

Other Categories

AI can also be categorized into subfields and approaches based on the tasks it aims to perform and its techniques. These categories, described in the following sections, provide an overview of the diverse areas within AI, each with its techniques, challenges, and applications. It should be noted that AI research continues to evolve, leading to new subfields and approaches as the field progresses.

Expert Systems

Expert systems, a type of AI program, excel at replicating human expertise within well-defined domains. They leverage extensive knowledge bases and rule-based reasoning mechanisms to offer insightful decisions and valuable recommendations, making them invaluable tools in fields like medicine and finance.

Fuzzy Logic

Fuzzy logic, a specialized field of artificial intelligence, is tailored to manage imprecise and uncertain data. Its applications are wide-ranging, especially in control systems and decision-making contexts, where establishing strict, binary rules is challenging. By accommodating ambiguity and uncertainty, fuzzy logic empowers AI systems to make more adaptable and context-aware choices, making it indispensable in real-world scenarios such as industrial automation and automotive control systems.

Swarm Intelligence

Inspired by the coordinated behavior of social insects such as ants and bees, swarm intelligence emulates decentralized and self-organized systems in AI. This approach finds applications in solving complex optimization and routing problems by harnessing the collective decision-making power of multiple agents. By simulating swarm behavior, AI systems can efficiently navigate intricate scenarios and find optimal solutions, making it particularly valuable in logistics, robotics, and other domains.

Machine Learning and Rule-Based Systems

Machine learning focuses on training algorithms to learn from data and make predictions or decisions. It includes techniques like deep learning, reinforcement learning, and supervised learning. On the other hand, Rule-based AI uses predefined rules and logic to make decisions. These rules are typically created by human experts and used in expert and knowledge-based systems.

Computer Vision

Computer vision is the field of AI dedicated to enabling computers to interpret and understand visual information from images and videos. It is used in image recognition, object detection, and facial recognition.

Robotics

AI in robotics revolves around crafting intelligent systems that can autonomously make decisions and engage physically with their surroundings. This encompasses a wide range of applications, from the development of self-driving cars that navigate real-world roads to automation in industrial settings, enhancing efficiency and safety.

Cognitive Computing

Cognitive computing systems strive to replicate human cognitive processes like learning, reasoning, problem-solving, and language comprehension. Notably, IBM's Watson is a prominent example of a cognitive computing system, demonstrating its capabilities in diverse domains, including healthcare and finance.

Appendix

AI Applications

Digital Assistants

Overview

AI has enabled the development of highly capable digital assistants that can perform various tasks and assist across various platforms and devices. Through advanced speech and language processing, AI transforms unstructured audio data into valuable insights and intelligence.

AI is powering a number of digital assistants, which use speech recognition to follow a user's command. These include Siri, Alexa, and Cortana. They collect information, interpret what is being asked, and then provide the answer. NLP can be used to process human speech and respond in the desired format, such as speech or text. Over time, these virtual assistants gradually improve and personalize solutions based on user preferences.

AI-powered digital assistants are continuously evolving and becoming more integrated into users' daily lives, offering greater convenience, accessibility, and assistance across a wide range of tasks and services.

A. Khan, *Artificial Intelligence: A Guide for Everyone*, https://doi.org/10.1007/978-3-031-56713-1

AI in Digital Assistants

AI-powered digital assistants have become an integral part of our daily lives, offering a wide range of functionalities that simplify tasks and enhance our interactions with technology. These versatile companions have found applications in numerous domains, contributing to increased productivity, accessibility, and convenience. Voice recognition and activation are among their fundamental features, enabling digital assistants to understand spoken commands and carry out tasks like setting reminders, sending messages, or playing music. By harnessing AI and voice recognition technology, digital assistants have revolutionized the way we interact with our devices and access information.

Information retrieval is another vital use case where AI plays a pivotal role. Digital assistants rely on AI algorithms to search the internet and promptly provide users with answers to questions, facts, definitions, and general knowledge queries. This information retrieval capability, coupled with AI's vast knowledge repository, transforms digital assistants into valuable sources of information at our fingertips. They empower users to quickly access data and facts, making them invaluable tools for both work and leisure.

AI-driven digital assistants have significantly streamlined task automation. These intelligent agents are responsible for automating routine tasks, such as sending emails, scheduling appointments, setting alarms, and even controlling smart home devices. By leveraging AI's capacity to understand user preferences and behavior, digital assistants deliver a more personalized and efficient task automation experience. This not only saves time but also enhances user convenience by reducing the burden of repetitive chores.

One of the notable advancements in the realm of digital assistants is their natural language understanding capabilities. AI enables digital assistants to comprehend and respond to natural language queries, allowing for more conversational interactions. This has transformed the way we interact with our devices, making conversations with digital assistants feel increasingly natural and intuitive. AI-driven digital assistants can engage users in dialogues, answer questions, and perform tasks through voice or text commands, revolutionizing user experiences.

Voice-controlled devices, like smart speakers and smartphones, rely on AI to offer hands-free control and information access through voice commands. The underlying AI technology powers these devices and connects them to a vast array of applications and services. Users can control their environments, access information, and perform various actions simply by speaking to these devices. Whether it is managing tasks, accessing information, or controlling smart home appliances, AI-driven digital assistants embedded in these devices play a central role in making our daily lives more efficient and enjoyable.

Personalized recommendations are another AI-driven feature that enhances user experiences. Digital assistants analyze user preferences and behaviors, harnessing the power of AI to provide personalized recommendations for music, movies, news, and other content. These recommendations are tailored to individual tastes and evolve over time, aligning with users' changing preferences. By leveraging AI's ability to process and understand user data, digital assistants deliver content that resonates with their audience.

Navigation and directions are areas where AI-powered digital assistants come to the rescue. By incorporating real-time data and AI algorithms, these assistants provide users with navigation assistance, helping them find the shortest routes, discover nearby places, and receive updates on traffic conditions. This use case is invaluable for travelers, commuters, and anyone seeking efficient and convenient ways to reach their destinations. AI-driven assistants offer real-time, location-based information that is both accurate and reliable.

Language translation is yet another application of AI integrated into digital assistants. These assistants can seamlessly translate spoken or written phrases into various languages by utilizing AI-based translation services. This functionality transcends geographical boundaries and language barriers, facilitating communication and making digital assistants valuable tools for individuals in multilingual environments. Users can rely on these digital companions to assist them in cross-cultural communication, whether for travel, business, or personal connections.

Health and wellness are domains where AI-powered digital assistants have made a substantial impact. These assistants track health metrics, set fitness goals, provide dietary advice, and remind users to take medications or engage in physical activities. They cater to the growing interest in personal health and well-being, offering support, guidance, and motivation. By drawing on AI's analytical capabilities, these digital assistants encourage users to lead healthier lifestyles, make informed health-related decisions, and stay on top of their well-being.

Accessibility features are yet another significant aspect of digital assistants. By providing voice-controlled accessibility features, such as voice-to-text and screen reader capabilities, digital assistants help individuals with disabilities access and interact with digital technologies. These features are vital for fostering inclusion, ensuring that everyone, including individuals with disabilities, can benefit from the advantages of AI and digital assistants, playing a crucial role in promoting accessibility and equal opportunities in the digital realm.

Content creation is an emerging application where AI-driven digital assistants prove their worth. These assistants help users draft emails, compose text messages, create documents, and even generate written content like articles or reports. They offer assistance by simplifying the writing process, suggesting content, and refining the language, saving users time and enhancing their productivity. By combining AI's language processing capabilities with the creative input of users, digital assistants open up new possibilities in content generation and text creation.

Smart home control is another dimension where AI-powered digital assistants have made substantial inroads. They connect to smart home devices and empower users to control lighting, thermostats, security systems, and other home appliances through voice commands. This use case epitomizes the concept of the IoT, where digital assistants act as central hubs for managing the myriad of connected devices that make up modern smart homes. This application adds an extra layer of convenience and automation to daily living, allowing users to easily control their home environments.

Conversational commerce is a growing trend facilitated by digital assistants. These assistants simplify the online shopping experience by guiding users in finding products, making purchases, and tracking orders through voice or text interactions. This innovative approach to e-commerce leverages AI to create a more intuitive and interactive shopping process. Users can ask questions, receive product recommendations, and complete transactions using conversational interactions, redefining the online retail landscape.

In emotional support, digital assistants offer more than just functional utility. Some digital assistants provide emotional support, engage in friendly conversations, and offer mental health resources to users in need. By recognizing and addressing users' emotional states, digital assistants extend their role beyond mere functionality, contributing to the well-being of individuals in times of stress, anxiety, or loneliness. These digital companions offer a comforting presence and valuable resources, highlighting the compassionate side of AI technology.

Calendar management is a practical use case where AI-driven digital assistants help users organize their schedules. They manage users' calendars, schedule appointments, and send reminders for meetings and events. By automating these tasks and keeping users informed, digital assistants enhance time management and reduce the risk of missing important engagements.

Voice banking and financial services are domains where digital assistants offer convenience and security. These assistants allow users to perform banking tasks, check account balances, and make financial transactions using voice commands and secure authentication. They add an extra layer of accessibility and efficiency to banking and financial interactions, reducing the need for manual input and enabling secure access to financial information.

In summary, AI-powered digital assistants have woven themselves into the fabric of our daily lives, offering a diverse array of applications that simplify tasks, enhance information access, and provide support across various domains. Their adaptability, personalization, and capacity to understand natural language have significantly altered the way we interact with technology. As digital assistants continue to evolve and expand their capabilities, they will likely find applications in new domains, further enriching our daily experiences and offering solutions to the ever-changing needs of modern life.

Self-Driving Cars

Overview

Artificial intelligence plays a crucial role in the development and operation of self-driving cars, also known as autonomous vehicles. A self-driving car is a vehicle that uses numerous sensors, radars, cameras, and AI to drive to destinations without needing a human driver. Examples include cars from Google and Tesla. The AI technology used for this application is computer vision and machine learning.

A self-driving car collects information from image recognition systems and neural networks, which distinguish patterns in the data fed to the AI calculations, including images from cameras. The neural networks figure out how to recognize traffic signals, trees, pedestrians, and other objects in the environment. AI simulates human perceptual and decision-making processes, using deep learning, and controls actions in driver control systems such as steering and brakes.

AI is central to the development of self-driving cars, enabling them to navigate complex environments, adapt to changing conditions, and make split-second decisions to ensure the safety and efficiency of transportation. As technology advances and AI algorithms improve, self-driving cars are expected to become more reliable and widespread in the future.

AI in Self-Driving Cars

The integration of artificial intelligence in self-driving cars has marked a transformative leap in the automotive industry, bringing us closer to the future of autonomous transportation. These AI-driven vehicles rely on a sophisticated network of algorithms and sensors to navigate and interact with their surroundings, ensuring safety, efficiency, and adaptability. The multifaceted role of AI in self-driving cars encompasses various crucial components that collectively create an autonomous driving experience.

Perception and sensor fusion are the foundational layers of AI in autonomous vehicles. These cars depend on a multitude of sensors, including LiDAR, radar, cameras, and ultrasonic sensors, to provide real-time data about their environment. AI algorithms process this data to generate a comprehensive understanding of the vehicle's surroundings. This enables the car to detect and identify objects, pedestrians, other vehicles, and road conditions with remarkable accuracy, forming the basis for safe and effective autonomous driving.

Object detection and classification is another fundamental aspect of AI in self-driving cars. These vehicles employ AI to detect and classify objects within their environment, differentiating between various entities like cars, cyclists, pedestrians, traffic signs, traffic lights, and obstacles. This capability is vital for ensuring the car's ability to make intelligent decisions and respond to different elements on the road, enhancing both safety and efficiency.

Mapping and localization constitute the infrastructure that supports self-driving cars. AI-driven mapping processes are responsible for creating and updating high-definition maps, which provide precise information about road geometry and lane markings. These maps are then utilized in conjunction with sensor data to help the vehicle determine its precise position on the road. The combination of mapping and localization is critical for the car's spatial awareness and effective navigation.

Path planning and decision-making are areas where AI plays a pivotal role in self-driving cars. AI algorithms assess real-time data, traffic conditions, and safety considerations to determine the optimal path and driving decisions for the vehicle. This includes making choices related to lane changes, merging onto highways, and navigating complex interchanges, demonstrating the adaptability and responsiveness of AI-driven autonomous vehicles.

Control systems are essential for executing planned maneuvers and maintaining safe and efficient driving. AI assumes control over the vehicle's acceleration, braking, and steering systems, ensuring that it adheres to the predetermined path and adheres to safe driving practices. AI-driven control systems are instrumental in translating decision-making algorithms into real-world actions, maintaining a high level of safety and precision.

Machine learning further enhances the capabilities of self-driving cars by enabling behavior prediction. These vehicles employ machine learning models to anticipate the actions of other road users, such as predicting when a pedestrian might cross the street or when another vehicle might change lanes. This predictive capacity is a cornerstone of ensuring the car's ability to respond proactively to dynamic traffic scenarios.

Sensor calibration and maintenance are critical aspects of AI's role in autonomous driving. AI helps monitor the health and calibration of sensors, ensuring they function correctly. When anomalies or issues are detected, AI can issue alerts for maintenance or cleaning, maintaining the reliability of the sensor suite.

In scenarios where a human is still involved, driver monitoring ensures safety. AI monitors the driver to detect signs of distraction or drowsiness. If such signs are detected, the AI system can alert the driver or, in some cases, take control of the vehicle to ensure safety.

Robustness and safety systems rely heavily on AI to respond appropriately to unexpected events, emergencies, and adverse weather conditions. AI helps create a safety net for the vehicle, enabling it to react swiftly and decisively when faced with unanticipated challenges.

Data collection and training are an ongoing process in the development of self-driving cars. These vehicles continuously collect data from sensors and road scenarios. This data is used to train and refine AI models, improving their performance and safety. The more data the AI system can gather and analyze, the more refined and effective it becomes.

Vehicle-to-everything (V2X) communication is an emerging trend that relies on AI to enhance safety and traffic management. V2X allows vehicles to communicate with each other and with traffic infrastructure. AI is critical in managing this complex web of communication, ensuring that data is shared and decisions are made in real-time to optimize safety and traffic flow.

Finally, simulation and testing are essential for the development of self-driving car algorithms and scenarios. AI-driven simulations create controlled environments in which various situations and scenarios can be tested before deploying them on public roads. This not only accelerates the development process but also ensures the safety and reliability of self-driving car systems.

In summary, the role of AI in self-driving cars is multifaceted and pivotal to the realization of autonomous transportation. AI components work together to provide a robust, adaptable, and safety-oriented driving experience. As technology continues to advance, AI will play an increasingly significant role in reshaping the future of mobility and enhancing safety, efficiency, and convenience on our roads.

Spam Email Filtering

Overview

AI is extensively used in spam email filtering to automatically identify and segregate unwanted or malicious emails from legitimate ones. An AI system scans each incoming message and flags or deletes it if it finds any objectionable content, such as spam or malware. Machine learning algorithms use statistical models to classify data. In the case of spam detection, a trained machine learning model is able to determine whether the sequence of words in an email is closer to those found in spam emails or safe ones. The process involves providing the machine learning model with a set of spam message examples, which lets it find the relevant patterns that separate spam from genuine messages.

The combination of AI-driven techniques enables email service providers and email clients to offer robust spam filtering solutions that can effectively block a significant portion of unwanted emails, including spam, phishing attempts, and malware-laden messages. While no system is perfect, AI significantly reduces the volume of unwanted emails that users have to sift through, enhancing email security and productivity.

AI in Spam Filtering

AI has revolutionized the way we combat spam emails, offering robust and efficient methods for filtering out unwanted and potentially harmful messages. The application of AI in spam filtering involves a multifaceted approach, with various components working in tandem to identify and quarantine spam. This holistic approach to spam filtering leverages AI algorithms and machine learning models, making it a dynamic and adaptive system capable of staying ahead of evolving spam tactics.

Feature extraction serves as the foundation of AI-driven spam filtering. AI algorithms meticulously dissect incoming emails, extracting a myriad of features. These features encompass a wide range of elements, including sender information, subject lines, the body text of emails, embedded links, and attachments. This comprehensive examination provides the AI system with a wealth of data to scrutinize, ensuring no potential spam indicators go unnoticed.

Machine learning models form the next layer of AI's arsenal in spam filtering. These models include both supervised and unsupervised algorithms, which are trained on vast datasets of labeled emails. This training process equips the models with the ability to recognize intricate patterns and characteristics that differentiate spam from legitimate messages. By ingesting and learning from a substantial volume of data, the machine-learning models develop a keen understanding of the hallmarks of spam, enabling them to make precise filtering decisions.

Content analysis is a crucial component of AI-powered spam filtering. AI algorithms dissect the text content of emails to uncover any suspicious keywords, phrases, or patterns, leveraging natural language processing techniques. The system actively seeks out linguistic characteristics commonly associated with spam messages. This content analysis extends beyond the mere presence of specific terms and delves into the context and usage of language within emails, enabling a deeper understanding of email content.

The evaluation of the sender's reputation is another critical function of AI in spam filtering. AI examines historical data to gauge the reputation of email senders. Senders who have previously been associated with spam or phishing emails are flagged as potential threats. This reputation-based approach provides a layer of defense against known malicious sources, reducing the likelihood of these messages reaching the recipient's inbox.

The behavioral analysis complements the automated scrutiny of email content. AI systems monitor user email behavior to detect anomalies. For instance, if a user begins to receive a sudden influx of emails from unknown senders or frequently opens suspicious attachments, the system may trigger alerts. This behavioral analysis enables proactive responses to deviations from a user's typical email interactions, enhancing security.

Pattern recognition is another vital facet of AI spam filtering. AI algorithms excel at identifying patterns within email content, which can signify various spam-related activities. These patterns include phishing attempts, deceptive subject lines, or the utilization of known spam email templates. By recognizing these recurrent patterns, the AI system can swiftly flag and quarantine potentially harmful emails.

Blacklists and whitelists serve as essential tools in AI-driven spam filtering. The system maintains lists of known spammers, often referred to as blacklists, and trusted senders, known as whitelists. When an email originates from a blacklisted source, it is automatically blocked or flagged as spam. Conversely, emails from whitelisted sources receive preferential treatment, bypassing strict scrutiny. This dual-list system streamlines the filtering process and contributes to accurate spam identification.

Real-time analysis is an integral aspect of AI spam filtering, ensuring that incoming emails are evaluated swiftly and efficiently. As emails arrive, AI algorithms assess them in real-time, comparing their characteristics to those of known spam. If an email exhibits attributes reminiscent of spam, it can be promptly flagged and quarantined before it reaches the recipient's inbox. Real-time analysis minimizes the potential for users to be exposed to spam messages.

Image analysis enhances spam-filtering capabilities, particularly when dealing with image-based spam. Some spam emails employ images with embedded text, a tactic intended to evade traditional content-based filters. AI employs optical character recognition (OCR) techniques to analyze these images, detecting any spam content concealed within them. By scrutinizing image content, AI ensures that image-based spam is not overlooked.

URL analysis is a vital component of AI spam filtering, as many spam emails contain URLs that lead to malicious websites. AI examines URLs within the email content, cross-referencing them with databases of known malicious links. Suspicious or blacklisted URLs are either blocked or flagged, preventing users from inadvertently accessing harmful websites. This proactive approach safeguards users from the risks associated with spam-related links.

User feedback plays a dynamic role in AI-powered spam filtering. AI systems consider user feedback, actively learning from user interactions with emails. When a user marks an email as spam, the system uses this feedback to improve its filtering accuracy. Over time, the system fine-tunes its algorithms based on user behavior, ensuring that the filtering process aligns with individual preferences and evolving spam trends.

Adaptive learning is a crucial element that enables AI spam filters to remain effective in the face of evolving spam techniques. As spammers continually devise new tactics, AI systems must adapt to stay ahead of emerging threats. Through continuous learning and model updates, AI spam filters remain at the forefront of spam detection and prevention, consistently improving their performance and efficiency.

In summary, AI's role in spam filtering is a multifaceted and dynamic process. The utilization of AI algorithms and machine learning models, combined with various analytical techniques, forms a comprehensive system capable of efficiently identifying and mitigating spam emails. The collaborative effort of these components ensures that spam filtering remains robust, proactive, and adaptable in the face of evolving threats, safeguarding users from unwanted and potentially harmful messages.

Social Media

Overview

AI plays a significant role in shaping the landscape of social media platforms, where it is widely used across all platforms, including Facebook, Instagram, Snapchat, Twitter, LinkedIn, and TikTok. It has transformed how brands are marketed and reduced costs across social media marketing activities by making them more efficient and automated. It has also enhanced user experiences, content moderation, advertising, and analytics.

AI has helped increase revenue because it learns from data and can use data from social media audiences to accelerate revenue in many ways. This includes learning which headlines, words, and images in posts lead to the most engagement, discovering new audiences and trends based on sentiment analysis, and predicting who will take action and buy more. It then reaches them with targeted social messages.

A key element of most companies' digital marketing strategy is posting and engaging on social media, which AI can help with. It can automate and scale the creation of content, automatically resize and reformat content for different channels, automatically target audiences with ads, and optimize advertising expenses.

With so much content being uploaded every minute, it is a real challenge for social media companies to moderate content and filter hate speech and violent content. AI has helped make this task manageable, compared to manual moderation which cannot adequately handle this task.

AI continues to evolve in the social media space, offering innovative solutions to enhance user experiences, support businesses, and improve content moderation. Its capabilities are expanding as social platforms harness AI's potential for greater customization, safety, and engagement.

AI in Social Media

AI's pervasive role in social media has ushered in a new era of personalized and efficient online experiences. Content recommendation stands out as one of the key applications of AI in this domain. AI algorithms meticulously analyze user behavior, encompassing actions such as likes, shares, and past interactions, to provide highly personalized content recommendations. This not only keeps users engaged but also encourages them to spend more time on the platform, ultimately benefiting content creators and advertisers. This tailored approach to content delivery has revolutionized how individuals interact with social media, aligning content with their preferences and interests.

Content moderation represents another vital area where AI solutions are indispensable. These content moderation tools act as vigilant gatekeepers, automatically detecting and filtering out inappropriate or harmful content. This includes identifying hate speech, spam, and graphic images ensuring a safer and more welcoming environment for users. By employing AI to handle content moderation, social media platforms are better equipped to uphold community standards and mitigate the risks associated with inappropriate content. This not only enhances the user experience but also safeguards the platform's reputation.

Chatbots and customer support services powered by AI have become a staple feature on social media platforms. These AI-driven chatbots adeptly respond to user inquiries, providing information, addressing common questions, and assisting with account-related issues. This not only offers users immediate assistance but also offloads routine customer support tasks from human agents, thereby improving efficiency. Chatbots showcase the power of AI in enhancing user engagement and experience on social media.

NLP represents another crucial facet of AI in social media. This technology empowers social media platforms to understand and analyze text-based content effectively. With NLP, platforms can identify trends, monitor sentiment, and track emerging topics in real-time. This not only allows for the quick identification of emerging conversations but also helps content creators and brands stay attuned to user sentiment and feedback. These insights enable businesses to adjust their strategies and refine content to better resonate with their target audience.

Sentiment analysis, a direct consequence of AI's language processing capabilities, holds a prominent place in the social media landscape. AI analyzes user-generated content to gauge sentiment, offering brands and marketers invaluable insights into public opinion and customer feedback. These insights, generated through AI-driven sentiment analysis, shape business strategies, product development, and communication approaches. By harnessing AI, brands and social media platforms can transform user-generated content into actionable information.

AI has significantly transformed the realm of advertising on social media. Ad targeting and personalization are powered by data derived from user profiles, behaviors, and preferences. AI leverages this information to deliver highly targeted advertisements, boosting the relevance of ads and increasing advertising ROI. With AI, users are presented with ads that align with their interests and needs, enhancing the user experience while providing businesses with a more effective advertising platform.

AI's capabilities extend to visual content as well. It enables image and video recognition, empowering automatic tagging, content recommendations, and accessibility features for visually impaired users. Recognizing objects, faces, and scenes in visual content allows for improved content organization, making it more accessible and engaging. It also provides new opportunities for accessibility, ensuring inclusivity on social media platforms.

Trend detection is another remarkable AI application in the social media landscape. AI constantly monitors social media conversations, swiftly identifying trending topics. Users and brands alike can stay updated on current events and discussions, making it a crucial tool for staying in tune with the ever-evolving landscape of social media.

Influencer marketing, a significant facet of social media advertising, also benefits from AI. AI thoroughly analyzes influencer profiles, engagement metrics, and audience demographics to identify the best influencers for collaboration on marketing campaigns. This data-driven approach to influencer selection ensures that brands can maximize the impact of their influencer partnerships.

AI's influence extends to content creation and enhancement. AI tools are adept at generating and enhancing visual content. They can produce photo filters, create captions, and offer design templates, streamlining the content creation process for users. This not only fosters creativity but also makes it easier for users to craft compelling posts, ultimately elevating the overall quality of content on social media platforms.

Real-time analytics, driven by AI, offer businesses valuable insights into user engagement, reach, and performance metrics. These insights help businesses make data-driven decisions, adapt their strategies on the fly, and stay responsive to changing market conditions and customer demands.

Automated social listening tools are another manifestation of AI's capabilities. They monitor conversations about brands and products across social media platforms, providing businesses with essential feedback for reputation management and strategy adjustments. This feature allows organizations to proactively address customer concerns and seize opportunities for improvement.

User segmentation is a key AI-driven application in social media. AI categorizes users into segments based on their behavior, demographics, and interests. This categorization allows businesses to tailor content and offers to specific groups, offering a more personalized experience and strengthening customer relationships.

Community management is another domain that benefits from AI. AI assists community managers in identifying and engaging with active and influential community members. Fostering a sense of belonging and loyalty plays a vital role in building vibrant and interactive communities on social media platforms.

Lastly, language translation services powered by AI bridge language barriers, enabling global interactions. AI's ability to translate content and comments into multiple languages broadens the horizons for communication on social media.

In summary, AI's applications in social media are extensive and transformative. From delivering personalized content recommendations to ensuring safety through content moderation, from improving customer support to analyzing sentiment and enhancing ad targeting, AI has revolutionized how social media platforms operate. The technology continues to shape the landscape of social media, making it more interactive, engaging, and personalized for users and more efficient and data-driven for businesses and content creators. The power of AI in social media is a testament to its versatility and potential to enhance the online experience.

Cybersecurity

Overview

AI-driven cybersecurity applications provide a formidable defense against the rapidly evolving digital threat landscape. AI has transformed the traditional cybersecurity paradigm, enabling proactive digital asset protection. These applications encompass a wide array of functions, including threat detection, analysis, behavioral analytics, anti-malware solutions, and identifying phishing attempts.

AI continuously monitors network traffic, scans for vulnerabilities, and enhances identity and access management (IAM) systems by evaluating user behavior. Real-time analysis of network traffic patterns and adaptive authentication are key strengths of AI in strengthening cybersecurity. Predictive analytics helps organizations stay ahead of potential threats, while security chatbots and automation streamline incident response. The relationship between AI and cybersecurity is symbiotic, with AI providing adaptive capabilities to match the ever-shifting threat landscape and cybersecurity relying on AI as a robust guardian of the digital age.

As the reliance on technology grows, so does the significance of safeguarding sensitive information from an ever-evolving landscape of cyber threats. In this dynamic cybersecurity paradigm, AI has emerged as a linchpin in fortifying defenses, detecting vulnerabilities, and responding swiftly to cyberattacks.

AI in Cybersecurity

One of AI's fundamental applications in cybersecurity lies in threat detection and analysis. AI-powered security systems stand vigilant, continuously monitoring network traffic, endpoints, and user behavior. The underlying machine learning algorithms, trained on vast datasets, identify subtle anomalies that might be indicative of a cyberattack. These patterns include indicators of malware, intrusions, or any activities deviating from the norm. By spotting such deviations in real-time, AI becomes the first line of defense in protecting organizations against malicious infiltrations.

Behavioral analysis is another critical domain where AI showcases its prowess. AI systems meticulously scrutinize user and entity behavior to establish a baseline of normal activity. When deviations occur, the AI systems promptly raise alerts. This behavioral analysis is invaluable for recognizing insider threats or compromised accounts. AI does not merely focus on external threats; it is equally attuned to the subtler dangers that could emanate from within an organization's own ranks.

AI-driven anti-malware solutions represent an important facet of cybersecurity. These solutions employ machine learning to identify and block new, previously unknown malware strains. They recognize these threats by discerning malicious patterns and behaviors in real-time. Phishing, a common vector for cyberattacks, is also in the sights of AI. AI algorithms analyze email content, sender behavior, and URL patterns to pinpoint phishing attempts, thereby shielding users from fraudulent emails and deceptive websites.

In an era where system vulnerabilities can be gateways for malicious actors, AI comes to the rescue through vulnerability assessment. AI tools thoroughly scan systems and applications to identify vulnerabilities and misconfigurations that attackers could exploit. It doesn't merely point out the problems but also helps prioritize patch management efforts, ensuring that limited resources are effectively channeled to mitigate the most critical risks.

User and entity behavior analytics (UEBA) leverages AI effectively in cybersecurity. These AI-driven platforms continuously monitor user and entity activities, especially in large and complex networks. Proficient at detecting abnormal behavior that may indicate insider threats, AI enhances security by recognizing patterns unnoticed by human operators. This capability strengthens an organization's ability to safeguard its digital assets, particularly in detecting insider threats. AI aids in preemptively identifying potential threats by closely monitoring user activity and promptly flagging suspicious behavior or data access. This proactive approach enhances an organization's trust and security from within.

AI extends its influence to the core of security information and event management (SIEM). These systems play a crucial role in modern cybersecurity by aggregating, correlating, and analyzing vast amounts of security data in real-time. Through AI's advanced analytical capabilities, SIEM solutions become more adept at detecting and responding to emerging threats. They offer a deeper level of insight into security incidents, enabling faster and more informed decision-making.

In an ever-connected world, where network traffic is the lifeblood of modern operations, AI's role in network traffic analysis is pivotal. AI is harnessed to scrutinize network traffic patterns, swiftly identifying any suspicious behavior. This helps in the early detection of intrusions and data exfiltration, where data is maliciously accessed or stolen from a network. AI's real-time analysis ensures that security breaches are spotted promptly and acted upon.

Identity and access management also benefits immensely from AI. AI bolsters IAM systems by continuously evaluating user access requests, identifying risky behaviors, and enforcing least privilege access policies. This is a fundamental step in safeguarding organizations against unauthorized access and data breaches, making IAM systems more robust and secure.

AI's capability extends to security chatbots that serve as indispensable assets to security teams. These AI-powered chatbots are versatile and can answer security-related inquiries, perform security assessments, and provide guidance during incidents. They function around the clock, offering valuable support and insights.

Adaptive authentication represents an intriguing dimension where AI plays a vital role. AI evaluates user behavior and context to determine the appropriate level of authentication required. By doing so, it reduces friction for legitimate users while simultaneously strengthening security against unauthorized access. The flexibility of adaptive authentication aligns security measures with actual risk, making the security framework more intelligent and user-friendly.

Predictive analytics is where AI demonstrates its potential for proactive cybersecurity. AI sifts through historical security data, identifying trends and patterns to predict future threats and vulnerabilities. By offering a glimpse into the future, organizations can be proactive, fortifying their defenses and staying one step ahead of potential attackers.

Security orchestration and automation have witnessed a revolution due to AI. AI-driven orchestration tools are capable of automating incident response and remediation processes. This reduces response times, ensures consistency in actions, and ultimately improves overall security. This confluence of AI and orchestration has redefined how organizations tackle cybersecurity incidents.

In summary, AI has become the backbone of modern cybersecurity, redefining how organizations safeguard their digital assets. Its influence covers real-time threat detection, behavioral analysis, anti-malware solutions, phishing detection, vulnerability assessment, network traffic analysis, SIEM systems, and endpoint protection. AI's role extends to user and entity behavior analytics, adaptive authentication, predictive analytics, insider threat management, security chatbots, and automated incident response. As the cyber threat landscape evolves, AI's adaptive capabilities ensure organizations remain resilient in the face of new challenges, making it the guardian of the digital age.

Language Translation

Overview

The landscape of language translation has undergone a profound transformation with the advent of AI technology. It has expedited the translation process and significantly elevated its accuracy and global accessibility. Unlike conventional machine translation, where words are directly converted from one language to another, AI delves deeper into linguistic nuances. It comprehends words, phrases, voice inflections, intricate sentence structures, and even colloquialisms, resulting in more precise and context-aware translation.

The impact of AI-driven language translation reverberates across diverse sectors, facilitating communication and transcending language barriers in realms such as business, travel, education, and diplomacy. As AI technology continues its rapid evolution, the future promises translations that are not only more accurate but also increasingly efficient and accessible to people worldwide. This synergy of AI and language translation not only breaks down linguistic boundaries but also bridges the gap between cultures and nations.

AI in Language Translation

Machine translation, driven by AI, has ushered in a new era of cross-linguistic communication. The emergence of neural machine translation (NMT) algorithms, as exemplified by platforms like Google Translate, has revolutionized the process of translating text from one language to another. Unlike earlier models, NMT systems, fortified by massive multilingual datasets, boast an impressive ability to comprehend and generate human-like translations. The underpinning principle of continuous learning empowers these AI translation models, enabling them to evolve and enhance their proficiency over time. This adaptability to new language patterns, colloquialisms, and evolving terminology fosters a dynamic and ever-improving translation process.

The multilingual support offered by AI-driven translation systems is nothing short of a linguistic marvel. Users can seamlessly translate text across a wide array of languages, ranging from the most widely spoken to the most obscure, thereby dismantling linguistic barriers that once stood as formidable roadblocks to global communication. Real-time translation capabilities have further fortified the arsenal of AI translation, facilitating instantaneous language conversion during voice or video conversations. This real-time prowess, applied in both personal and professional contexts, has become a driving force behind breaking down language barriers in an increasingly interconnected world.

Contextual understanding, a hallmark of AI translation, significantly elevates the quality and precision of translated content. These AI models do not operate in isolation; instead, they meticulously consider the broader context of the text they are translating, enabling a customized experience. This contextual awareness allows them to analyze surrounding words and phrases, crafting translations that are not just linguistically accurate but also contextually relevant.

Moreover, customization is a hallmark feature of AI translation systems, offering users the power to tailor their translation preferences. Whether choosing between formal or informal language styles, regional dialects, or industry-specific terminologies, AI-driven translation provides an experience attuned to the user"s unique requirements. In the field of document translation, these systems excel at translating entire documents, encompassing diverse formats like contracts, articles, reports, and more, all while preserving the original formatting and layout.

The digital landscape is replete with websites, and AI-driven website translation tools are invaluable in ensuring that content is accessible to a global audience. By automating the translation process, these tools circumvent the need for manual intervention in the localization of web content. Speech translation is another remarkable facet of AI translation. Leveraging speech recognition and translation systems, AI can convert spoken language into text and then translate it into the desired language. This real-time speech translation has manifold applications, from international conferences and travel to providing multilingual customer support services.

Quality assurance is a critical component in the translation process, and AI plays an essential role in identifying and rectifying translation errors. AI acts as a quality gatekeeper by ensuring the accuracy and fluency of translated content. AI extends its expertise to content localization, enabling the adaptation of material for specific regions or cultures. This encompasses the adjustment of date formats, currency, cultural references, and other contextual elements.

AI also plays a pivotal role in enhancing the user experience by ensuring that multimedia content, including videos and podcasts, remains inclusive and comprehensible for a global audience, as it expertly converts spoken words into text and provides translations in multiple languages through subtitles and captions.

Translation memory, another AI-driven tool, significantly streamlines translation workflows by storing and reusing previously translated phrases and sentences. This approach not only expedites the translation process but also helps reduce the associated time and costs.

AI translation transcends mere convenience and efficiency; its transformative power lies in its ability to break down language barriers and democratize access to knowledge and information, ensuring that even those with limited language proficiency can engage with a wealth of content and resources, thereby fostering greater inclusivity and understanding across the global community.

The significance of AI translation extends into the sphere of e-commerce and global business. For international companies, AI translation plays a pivotal role in marketing products and services to diverse audiences, managing international customer support, and facilitating expansion into new markets. In this context, AI is more than a translation tool; it becomes a key enabler of global business strategies.

In summary, AI has fundamentally reshaped language translation, making it faster, more accurate, and universally accessible. Continuous learning, multilingual support, real-time translation, contextual understanding, customization, and the ability to handle diverse content types have propelled AI-driven machine translation to the forefront of cross-linguistic communication. Its applications span from individual users seeking seamless communication to global businesses looking to expand their reach and market their products to a diverse and multicultural audience. With AI at the helm, language translation has transcended linguistic boundaries, promoted worldwide connections, and fostered understanding among people from different corners of the world. As AI technology continues to advance, the future promises even more sophisticated, efficient, and inclusive translations, further fortifying the bridge between languages and cultures.

Forecasting Demand

Overview

AI-powered demand forecasting is revolutionizing the way businesses anticipate consumer needs and make strategic decisions. At its core, AI-driven demand forecasting involves collecting, analyzing, and integrating vast datasets from diverse sources, ensuring a comprehensive understanding of market dynamics. These systems utilize advanced techniques such as time series analysis, machine learning algorithms, and statistical methods to unveil hidden patterns, trends, and seasonality in historical data. AI's adaptability allows it to continuously learn and improve over time, enhancing its forecasting accuracy by adapting to new language patterns, slang, and emerging terminology.

AI further extends its capabilities to real-time data analysis, making it an invaluable asset for businesses in a rapidly changing, interconnected world. By harnessing the potential of AI, businesses can access more accurate, efficient, and real-time demand forecasting, optimizing operations and boosting adaptability in a dynamic marketplace.

AI in Demand Forecasting

Demand forecasting has witnessed a profound transformation with the integration of AI. At the heart of AI-driven demand forecasting lies a complex interplay of data collection and integration. These systems amass vast troves of data from a myriad of sources, encompassing historical sales records, customer orders, website traffic, social media mentions, and economic indicators. This comprehensive data collection, a fundamental facet of AI's capabilities, sets the stage for more accurate analyses. However, the process does not end with data aggregation; AI goes a step further by engaging in meticulous data cleaning and preprocessing. This includes identifying and rectifying inconsistencies, addressing missing values, and managing outliers. The goal is to ensure that the data underpinning the forecasts is of the highest quality, a critical prerequisite for precision in demand forecasting.

Time series analysis is another pivotal dimension where AI plays a decisive role. Employing sophisticated techniques, AI scrutinizes historical data, unearthing hidden trends, recognizing patterns of seasonality, and identifying recurring behaviors in demand. This historical analysis serves as a crystal ball for forecasting future demand, offering valuable insights into consumer behavior and preferences.

AI's repertoire is not confined to mere data analysis; it extends to the realm of machine learning algorithms. It uses algorithms like linear regression, exponential smoothing, and neural networks to improve forecasting accuracy through adaptability and learning mechanisms. These algorithms are not static; they evolve with data, making them more robust and proficient over time.

Furthermore, AI harnesses advanced statistical methods such as autoregressive integrated moving average (ARIMA) and Prophet, a forecasting tool developed by Facebook, to unravel complex demand patterns and variations. The fusion of AI and advanced statistics enables businesses to grasp the intricacies of demand behavior, no matter how intricate or multifaceted.

Market segmentation, an indispensable component of demand forecasting, is another area where AI shines. By segmenting markets and customer groups based on various factors like geography, demographics, or purchase history, AI tailors forecasts for different product categories or customer segments. This granular approach to forecasting empowers businesses to make more precise decisions in terms of inventory management, marketing, and product development.

Demand forecasting does not exist in a vacuum. It must take into account external influences, and AI readily incorporates external data sources to ensure a comprehensive perspective. Factors like weather forecasts, holidays, and economic indicators, when factored into forecasting models, allow businesses to better grasp how external forces may impact demand. This makes the forecasts not just accurate but also resilient to external contingencies.

In today's fast-paced business landscape, demand is anything but static. AI is equipped to handle this dynamism through a technique known as demand sensing. These systems continuously monitor real-time data streams, making it possible to respond rapidly to sudden changes in demand or emerging trends. This agility in demand forecasting is indispensable for staying competitive and agile in a rapidly evolving marketplace.

Forecast accuracy evaluation is an essential step in the AI-driven demand forecasting process. AI creates a feedback loop that continually refines forecasting models by comparing predicted demand to actual sales data. This perpetual process of learning from its own forecasts contributes to ever-increasing accuracy.

The power of AI is magnified when it comes to scenario analysis. AI empowers businesses to simulate different scenarios and evaluate the potential impact on demand. Whether it is assessing the consequences of changing market conditions or pricing strategies, scenario analysis plays a pivotal role in informed decision-making and risk management.

AI does not merely stop at forecasting; it extends its capabilities to demand shaping. By recommending pricing changes, promotions, and inventory management techniques, AI assists businesses in aligning supply with expected demand. This demand shaping is instrumental in not only maximizing profits but also ensuring a seamless customer experience.

Collaborative forecasting is yet another facet of AI's multifaceted role in demand forecasting. AI fosters collaboration by providing a shared platform for departments such as sales, marketing, and supply chain. The outcome is a more holistic approach to forecasting, where different facets of the business converge to create a more comprehensive and aligned forecast.

Inventory optimization is a natural byproduct of AI-driven demand forecasts. By providing businesses with a clearer picture of expected demand, AI helps optimize inventory levels, reducing excess inventory costs and ensuring that products are available when needed. The concept of just-in-time manufacturing, which is predicated on producing products in line with actual demand, has also been significantly bolstered by AI's forecasting capabilities. This minimizes waste and storage costs, offering a win-win scenario for businesses and the environment.

The integration of AI-driven demand forecasts into supply chain management is a pivotal step in ensuring efficiency. AI optimizes logistics, transportation, and distribution by aligning them with demand forecasts, streamlining the supply chain. This not only results in cost savings but also ensures that products reach customers on time, enhancing customer satisfaction and brand reputation.

In summary, AI has transformed demand forecasting, enhancing precision, adaptability, and efficiency. By collecting, preprocessing, and analyzing diverse data sources, it enables more accurate forecasts. Utilizing machine learning algorithms, statistical techniques, and external data sources provides a nuanced understanding of demand, accounting for complexity and external influences. AI goes beyond forecasting, extending its capabilities to optimize demand, fostering better decisions and resource allocation. By enhancing collaboration among different departments and improving inventory management, AI-driven demand forecasting fundamentally changes how businesses operate, becoming essential for competitiveness in a dynamic marketplace.

Manufacturing

Overview

AI is transforming the manufacturing industry by optimizing processes, increasing efficiency, and enhancing product quality. Manufacturing areas in which AI is being used include adapting to uncertain and variable supply and demand, forecasting demand, ordering materials, scheduling, logistics, predicting and preventing machine downtime, streamlining processes, quality control, reducing waste, and optimizing inventory management.

AI-driven manufacturing not only enhances productivity but also contributes to sustainability and flexibility in production. As AI technology continues to advance, manufacturers can achieve higher levels of automation, precision, and adaptability, making them more competitive in today's global markets.

AI in Manufacturing

Production optimization is an area where AI excels. AI algorithms can autonomously optimize production processes by adjusting real-time parameters such as temperature, pressure, and speed. This adaptability ensures that manufacturing operations are precisely aligned with desired outcomes, enhancing product consistency while reducing resource wastage. It is a win-win scenario that modern industry cannot ignore.

Quality control and inspection in manufacturing processes are paramount to ensuring product excellence. AI-driven computer vision systems have taken the lead in this domain. These systems, equipped with the power of AI, scrutinize products for even the minutest defects, upholding the gold standard for manufacturing quality. By identifying minor imperfections and anomalies in real-time, AI-powered inspection systems contribute to error reduction and yield consistency.

Process control is a critical aspect of manufacturing, where AI assumes a pivotal role. It meticulously monitors and controls complex manufacturing processes, making real-time adjustments to maintain product quality and efficiency. This proactive approach to process management ensures that products meet stringent quality standards consistently.

AI has benefited manufacturing by reshaping supply chains through demand forecasting, inventory optimization, and the swift identification of logistics bottlenecks. This transformation results in reduced lead times, lower operational costs, and an efficient supply chain. Consequently, industries are better equipped to meet customer expectations and market demands. AI also extends its influence to supply chain visibility, empowering manufacturers with real-time analytics, facilitating swift responses to disruptions, and supporting informed decision-making. In the dynamic landscape of modern industry, AI-driven analytics ensure adaptability and agility.

The advent of AI-driven robots and robotic arms has brought forth a new era in automation. These robotic entities work with precision and speed, undertaking tasks such as assembly, welding, and material handling, offering new possibilities for manufacturers to increase their productivity. Moreover, collaborative robots are AI-powered machines designed to work safely alongside human workers. They bolster manufacturing efficiency, particularly in tasks requiring precision and strength, underscoring the harmony of human–machine collaboration in modern industry.

AI has also influenced product design through AI-driven design tools that generate product designs, optimizing manufacturing efficiency, materials, and cost. This allows manufacturers to harness AI to create inherently efficient designs for production. As customization and personalization have become cornerstones of modern manufacturing, AI plays a central role in achieving mass customization. It facilitates the production of highly customized products at scale, addressing individual customer preferences and demands with remarkable agility.

Material selection, waste reduction, and inventory management are critical concerns for sustainable manufacturing. AI's data-driven approach helps in choosing the right materials for production, reducing waste, and ensuring that manufacturing practices remain eco-friendly and efficient. Additionally, AI extends its reach to inventory management, where it optimizes inventory levels. This involves minimizing excess stock while ensuring products are available when needed, avoiding the costs and inefficiencies associated with overstocking.

Demand forecasting is another forte of AI. By analyzing historical data and market trends, AI predicts future demand, aiding in production planning and inventory management. Industries now benefit from greater precision and adaptability in addressing market fluctuations and customer needs.

Maintenance scheduling with AI is not just about keeping the machinery running; it is about minimizing disruptions. AI schedules equipment maintenance during periods of low production demand, optimizing operational continuity while ensuring maintenance is conducted when it is least disruptive. AI also contributes to root cause analysis in manufacturing by identifying the underlying reasons for manufacturing issues. This, in turn, aids organizations in addressing root causes and preventing their recurrence, thus enhancing overall operational efficiency.

AI shines in energy management as well. AI optimizes energy consumption in manufacturing facilities by analyzing sensor data and adjusting usage patterns. This translates to reduced operational costs and a reduced environmental footprint, aligning industries with sustainable and cost-effective energy practices.

Human–machine collaboration is pivotal to modern manufacturing, where AI collaborates with human workers, providing them with data, insights, and support to enhance decision-making and productivity. This combination of human ingenuity and AI-driven efficiency yields exceptional results. Moreover, worker safety is of utmost importance in manufacturing, and AI plays a pivotal role in ensuring a secure work environment. It monitors the work environment to ensure safety compliance and detects potential hazards, protecting workers from harm and preventing accidents.

In summary, AI's influence spans predictive maintenance, quality control, production optimization, supply chain management, collaborative robotics, process control, energy management, supply chain visibility, customization, material selection, waste reduction, inventory management, demand forecasting, maintenance scheduling, product design, and human–machine collaboration. As AI advances, manufacturing stands on the brink of an exciting future, promising greater efficiency, reduced costs, and a commitment to sustainability. AI is reshaping the manufacturing landscape, setting new standards for quality, efficiency, and adaptability.

Predictive Maintenance

Overview

Predictive maintenance, leveraging the transformative capabilities of AI, emerges as a leading practice in modern industrial operations. Across industries, particularly in manufacturing, AI plays a pivotal role in enhancing operational efficiency, cutting costs, and ensuring top-tier quality standards. This proactive application of AI involves analyzing data from sensors and machines to predict equipment failures before they occur, revolutionizing maintenance strategies. By adopting a proactive stance, predictive maintenance, powered by AI algorithms, not only reduces downtime and maintenance costs but also elevates equipment reliability and extends its lifespan.

This shift toward a data-driven and proactive approach represents a fundamental change, enabling organizations to enhance overall operational efficiency and productivity. Harnessing the power of artificial intelligence allows companies to predict failures, optimize maintenance processes, and significantly improve asset reliability, thereby shaping the future of industrial maintenance practices.

AI in Predictive Maintenance

Predictive maintenance relies on robust data collection and sensor integration, with AI playing a pivotal role in ensuring optimal equipment reliability. AI systems gather data from various sensors and IoT devices installed on machinery, monitoring factors like temperature, vibration, pressure, and performance metrics. This data serves as the basis for predictive maintenance strategies. To make this data useful, AI takes on the critical task of data preprocessing. This process involves cleaning the data, identifying outliers, and handling missing values. Clean and reliable data is essential for accurate predictions, making data preprocessing an indispensable function within the predictive maintenance framework.

Feature engineering is a crucial process where engineers and data scientists select relevant features, which are data attributes, to feed into the predictive models. Feature engineering helps identify the most indicative factors of impending equipment failures, ensuring that models receive the most pertinent information. AI leverages machine learning techniques, including regression, classification, and time-series analysis, to construct predictive models. These models learn from historical data patterns and extrapolate them into forecasts of equipment failures.

Anomaly detection is integral to predictive maintenance, with AI-based algorithms identifying deviations from normal equipment behavior. Maintenance alerts are triggered when an anomaly is detected, indicating potential equipment failures. Predictive models forecast the remaining useful life of equipment or estimate the probability of failure within a certain timeframe, ranging from simple threshold-based indicators to more complex predictive algorithms.

AI continuously monitors real-time sensor data, comparing it to predictive models. The system generates maintenance notifications when a deviation from predicted behavior is detected. This real-time monitoring is instrumental in addressing potential failures preemptively. Beyond predicting failures, AI also recommends specific maintenance actions. For instance, AI can suggest component replacements or specific maintenance tasks, streamlining the maintenance process with precision. AI-based predictive maintenance encourages a shift from fixed maintenance schedules to condition-based maintenance, where equipment is serviced when needed, reducing costs and minimizing downtime.

AI excels in historical data analysis, identifying patterns and trends that help organizations refine maintenance strategies over time, optimizing the entire maintenance process. Since AI provides deep insights into equipment performance, it allows organizations to make data-driven decisions about maintenance, upgrades, and replacements. It also supports comprehensive asset lifecycle management, tracking equipment from procurement to retirement, empowering organizations to maximize asset value throughout their lifecycle.

In summary, AI in predictive maintenance has ushered in a new era of equipment management. By seamlessly integrating data collection, preprocessing, predictive modeling, and real-time monitoring, AI ensures the reliability and efficiency of critical machinery and equipment. This not only reduces operational disruptions but also saves costs and maximizes asset value. Predictive maintenance, driven by AI, is poised to become a linchpin of efficient asset management in the modern era.

Supply Chain

Overview

AI has revolutionized supply chain management, offering advanced tools and techniques that enhance efficiency, visibility, and decision-making. Through the analysis of various datasets, AI enables the optimization of supply chains across different stages, including predicting product demand, inventory management, sourcing, shipping, sales and consumer trend predictions, demand forecasting, and capacity planning. This comprehensive approach leads to higher productivity, quality improvement, increased output, and helps in avoiding supply chain disruptions.

Organizations leveraging AI technologies in their supply chain operations gain greater agility, responsiveness, and cost-effectiveness. AI-driven insights and automation empower supply chain professionals to make informed decisions, proactively address challenges, and enhance customer satisfaction and competitiveness, transforming logistics, inventory, and distribution processes. AI plays a multifaceted role in optimizing and streamlining supply chain operations.

AI in the Supply Chain

One of the pivotal applications of AI in supply chain management is demand forecasting. AI models delve into historical data, market trends, and external factors to generate highly accurate demand forecasts. This allows organizations to make informed decisions about inventory levels, minimizing stockouts while avoiding overstocking, ultimately reducing carrying costs and enhancing overall operational efficiency.

Supply chain planning, a complex and multifaceted task, greatly benefits from AI. The technology optimizes various constraints and variables to determine the most efficient production schedules, distribution routes, and sourcing strategies. By balancing costs and service levels, AI enhances the planning process, ensuring resources are allocated effectively, and inefficiencies are reduced.

Route optimization is an area where AI algorithms make a substantial impact. They carefully calculate the most efficient transportation routes for deliveries and shipments. These calculations take into account real-time data such as traffic conditions, weather, and fuel costs. The result is reduced transportation expenses and improved delivery accuracy, enabling organizations to provide more efficient and cost-effective services to their customers.

AI also supports sustainability and green initiatives in supply chain operations. By optimizing transportation routes and reducing energy consumption, organizations can minimize their environmental footprint and support sustainability goals. Last-mile delivery optimization is another significant area for AI applications. By considering factors such as delivery windows, traffic patterns, and customer preferences, AI optimizes the final leg of the supply chain, resulting in more efficient and customer-centric delivery services.

Inventory management is another domain where AI shines. These systems continuously monitor inventory levels in real-time and automatically trigger reorder points. By considering critical factors such as lead times, demand fluctuations, and supplier performance, they optimize inventory management.

This results in a leaner supply chain with reduced carrying costs and improved responsiveness to changes in demand.

Warehouse automation is another arena where AI has proven invaluable. AI-powered robots and automation systems enhance warehouse operations by autonomously handling tasks like picking, packing, and inventory movement. This automation improves efficiency, reduces labor costs, and minimizes errors, ensuring that warehouses operate smoothly.

Quality control is another crucial aspect of supply chain operations that benefits from AI. Computer vision powered by AI can inspect products and packaging during the manufacturing and distribution processes, detecting defects with precision. This not only ensures product quality but also reduces waste and associated costs.

Supplier management is enhanced through AI's ability to track supplier performance. Metrics such as delivery times, product quality, and pricing are closely monitored. AI helps organizations identify potential issues and opportunities for improvement, ultimately leading to stronger and more reliable supplier relationships. Supplier collaboration is made more efficient through AI, with real-time communication and data-sharing capabilities streamlining procurement processes and reducing lead times.

Order fulfillment processes are optimized through AI's orchestration of tasks. The technology prioritizes orders, efficiently allocates inventory, and coordinates warehouse and distribution center workflows. This results in a streamlined and efficient fulfillment process.

Customer service is another domain where chatbots and virtual assistants come into play. These virtual agents assist customers with order inquiries, delivery status, and returns, reducing support costs and improving the customer experience.

Cold chain monitoring is essential for ensuring the integrity of perishable goods. AI and IoT devices play a vital role in monitoring and controlling temperature-sensitive shipments, guaranteeing that products reach their destinations in optimal condition.

AI also has a role in risk management. AI assesses and mitigates supply chain risks by analyzing various risk factors like geopolitical events, weather disruptions, and market volatility. This proactive approach allows organizations to respond effectively to potential disruptions and minimize their impact.

Finally, AI offers insights into supply chain performance through data analytics, enabling organizations to identify bottlenecks, inefficiencies, and areas for improvement. AI-driven continuous improvement ensures that supply chain operations remain agile and responsive to changing market conditions and customer demands.

In summary, AI's integration into supply chain operations has revolutionized the way businesses manage their logistics, inventory, and distribution processes, enhancing efficiency, reducing costs, and providing better service to their customers, ultimately driving success in an increasingly competitive market.

Video Analytics

Overview

AI plays a crucial role in video analytics, transforming the way it analyzes and derives insights from video data. Computer vision AI systems focus on analyzing digital images, videos, and other visual inputs. Such systems can be used for video analytics and surveillance in a variety of workloads and operating conditions. AI can be integrated with cameras, enabling the real-time analysis of video content, extraction of metadata, the sending of alerts, and providing actionable intelligence to security personnel or other systems.

AI-driven video analytics enhances efficiency, safety, and decision-making in various industries and applications. It enables organizations to process and analyze vast amounts of video data in real-time, providing valuable insights and actionable information. In an age where visual data is omnipresent, the role of AI in video analytics stands as a transformative force, reshaping how we understand and interact with the world.

AI in Video Analytics

Object detection and tracking are at the forefront of this revolution, as AI algorithms have evolved to seamlessly identify and monitor objects, people, vehicles, and even animals in the dynamic tapestry of video streams. The applications of this technology are as varied as the imagination itself, finding utility in domains as diverse as surveillance, traffic management, and crowd monitoring, where the need for real-time understanding and action is paramount.

AI steps onto the roads with license plate recognition systems as well. These AI-driven systems deftly read and record license plates from video feeds, ushering in more efficient parking management toll collection and aiding law enforcement in identifying vehicles swiftly and accurately. In traffic management, AI proves to be a game changer. Real-time analysis of traffic patterns, congestion, and accidents empowers AI systems to make critical recommendations and decisions, leading to more efficient traffic management and enhanced road safety.

AI's grasp extends to object counting and density estimation, a technology that holds significance for capacity planning in public spaces and retail stores. Its ability to count objects or people within video frames and estimate crowd density brings a new level of accuracy and efficiency in managing spaces and resources. Video summarization is yet another dimension of AI's prowess, simplifying video review and analysis by condensing lengthy video footage into shorter clips, highlighting key events or moments, thus facilitating swifter insight extraction.

AI-driven video analytics extends its capabilities with anomaly detection, identifying unusual or suspicious behavior by comparing it to predefined patterns. This functionality acts as an ever-watchful guardian, flagging unauthorized access to secure areas and detecting peculiar crowd behavior in public spaces. Beyond security, these applications have implications that go beyond, where early recognition of irregularities is pivotal. In the realm of video analytics, security and intrusion detection exemplify AI's prowess. Engineered to identify potential security threats like intruders or suspicious behavior in real-time, these systems usher in a new era of security where threats prompt proactive responses.

Another facet showcasing AI's capabilities lies in facial recognition, providing real-time identification and verification of individuals, especially within the realm of video analytics. Its applications extend across access control, identity verification, and security, where the ability to recognize and verify individuals quickly and accurately is indispensable. The emotional landscape also falls under the purview of AI, as it delves into emotion analysis by examining facial expressions and gestures, particularly relevant in video analytics applications. This technology is instrumental in marketing, customer feedback analysis, and the security domain, where understanding the emotional state of individuals holds paramount importance in the context of video analytics.

AI's capabilities are not limited to visual data alone; they also encompass the auditory. The technology's speech and audio analysis proficiency includes transcribing speech, detecting keywords, and identifying audio anomalies. This versatility is paramount for transcription services and audio surveillance, where deciphering and interpreting audio content is pivotal.

The industrial sector is not left untouched, with predictive maintenance emerging as a frontrunner. AI analyzes video feeds from machinery to detect the early signs of wear and tear or potential equipment failures, facilitating predictive maintenance strategies, minimizing downtime, and reducing maintenance costs significantly.

In the digital sphere, AI finds itself tasked with yet another mission—content moderation. It is instrumental in automatically moderating user-generated video content on platforms, vigilantly identifying and removing inappropriate or prohibited content, and ensuring safe and healthy online environments.

The environment, too, is under AI's watchful gaze as it analyzes video data from environmental sensors and cameras to monitor and respond to events like wildfires, floods, and pollution, ensuring that our planet's health is monitored and protected. Additionally, the healthcare and medical imaging sectors benefit immensely from AI applications. These include monitoring patient vital signs, analyzing surgical procedures, and detecting anomalies in medical images, thereby redefining diagnostic accuracy and healthcare precision.

At the crossroads of data and shopping, retail analytics leverages AI-powered video analytics to understand customer behavior, optimize store layouts, and analyze shopping patterns, ultimately enhancing the retail experience and business efficiency. Also, the realm of sports is no stranger to AI's transformative influence. AI is harnessed in the realm of sports for player tracking, performance analysis, and the real-time generation of statistics.

In summary, AI's foray into video analytics marks a pivotal moment in our technological evolution, forever altering how we perceive, understand, and respond to the visual data surrounding us. From enhancing security and safety to redefining efficiency in retail, healthcare, and the environment, AI's capabilities are ever-expanding. It holds the promise of a future where we stand better equipped to harness the power of visual data to improve lives, foster efficiency, and enrich our understanding of the world.

Human Resources

Overview

AI is revolutionizing the field of human resources (HR) by optimizing processes, boosting decision-making, and elevating the overall employee experience. The integration of AI with HR practices has ushered in numerous benefits across the human resources spectrum. It has remarkably enhanced talent acquisition and recruitment processes by streamlining hiring procedures, removing biases in candidate selection, and accelerating the screening of resumes to shortlist the most suitable candidates. This not only reduces the hiring cycle time but also results in more efficient onboarding of new employees.

AI-driven HR is not only about efficiency but also data-driven decisions. It involves collecting employee feedback, predicting their concerns, and addressing issues more effectively. This approach translates into faster and better decision-making across various HR functions. Moreover, AI aids in matching employee competencies with project assignments, thereby ensuring that teams are built with the right skills, improving overall productivity. In the ever-evolving landscape of HR, AI is a powerful ally that is reshaping the HR profession and, in turn, the employee experience. As AI continues to evolve, its role in HR is expected to expand, promising even more strategic and transformative HR practices.

The synergy between AI and HR underscores a fundamental shift in how organizations manage their most valuable resource—human capital. By enhancing recruitment, talent management, and employee development, AI empowers HR professionals to focus on more strategic and impactful initiatives, ultimately leading to a more efficient and satisfied workforce.

AI in HR

The recruitment and talent acquisition process has seen a remarkable evolution through AI-powered tools. From swiftly analyzing resumes to identifying relevant skills, AI can efficiently match candidates with job openings, streamlining the hiring process. Chatbots and virtual assistants now handle initial candidate inquiries and even schedule interviews, providing prospective employees with quick and seamless interactions from day one.

AI's transformative reach extends further with resume screening. AI algorithms can quickly and accurately review and screen a vast volume of resumes. Shortlisting candidates based on predefined criteria offers a significant time-saving advantage to HR professionals, allowing them to focus on more strategic aspects of their roles. Candidate assessment also benefits immensely from AI. AI-driven assessment tools provide objective insights into candidates' capabilities through various methods, including video interviews, coding challenges, and psychometric tests. This ensures that the hiring process is based on merits, ultimately enhancing the quality of new hires.

Diversity and inclusion have gained from AI's impartial approach. AI fosters fairer hiring practices by removing biases from job descriptions and assessing candidates purely on their qualifications and skills. It actively contributes to promoting diversity within organizations, ensuring that talent is the prime focus during recruitment.

Employee onboarding, often a critical phase in the employee lifecycle, is made more engaging and efficient through AI-driven systems. These systems provide new hires with essential information, training modules, and answers to common questions. This streamlined process sets a positive tone for employees' experiences from their very first day. AI is also a force in enhancing employee engagement and feedback. AI-powered surveys and sentiment analysis tools collect and analyze employee feedback, providing HR departments with valuable insights into employee sentiments and areas for improvement. The result is a more engaged and satisfied workforce.

Performance management is yet another area where AI plays a significant role. It assists in setting performance goals, tracking progress, and providing recommendations for employee development. With AI, performance appraisals become more accurate and data-driven, facilitating more meaningful discussions between employees and their managers. The learning and development domain benefits from AI's ability to recommend personalized training programs for employees based on their roles, skills, and career goals. AI empowers employees with the resources they need to grow professionally, contributing to their career satisfaction and overall productivity.

Predictive analytics for retention is a game-changing application of AI. By analyzing historical data, AI can predict which employees are at risk of leaving the organization. HR can then take proactive measures to retain valuable talent, ultimately reducing turnover rates and the associated costs. AI's influence extends to workforce planning, enabling HR departments to forecast future workforce needs. This includes hiring, reskilling, and restructuring based on business goals and market trends, ultimately fostering a more agile and responsive organization.

AI also simplifies employee benefits management by analyzing individual needs, family size, and preferences. This helps employees choose the best benefits packages, ensuring their choices suit their unique circumstances well. The advent of chatbots for HR support marks a new era of employee assistance. Chatbots are available 24/7 to provide employees with quick answers to HR-related questions, whether it is regarding benefits information, leave requests, or policy inquiries. This technology simplifies HR interactions, making employee support more accessible and efficient.

In the realm of payroll and compensation management, AI's role cannot be overstated. By automating payroll calculations, AI ensures accuracy and compliance with tax regulations. Moreover, it assists in determining competitive compensation rates, contributing to fair and equitable compensation practices within the organization.

Employee well-being is paramount, and AI plays a critical role in ensuring it. By monitoring factors such as workload, stress levels, and work–life balance, AI enables HR to intervene, when necessary, ultimately creating a healthier and more supportive work environment. AI is also a guardian of ethics and compliance, as AI-driven systems enable employees to report compliance violations and ethical concerns confidentially. This fosters a culture of transparency, ensuring that organizations uphold their ethical standards.

Exit interviews and offboarding procedures are streamlined through AI-driven surveys and analysis tools. These tools gather feedback from departing employees, helping HR identify trends and areas for improvement, ultimately leading to a more refined and employee-focused exit process.

In summary, AI's integration with HR marks a significant step forward in the evolution of human resource management. AI's contributions are far-reaching, from the hiring process to employee development, retention, and beyond. By streamlining HR processes, fostering diversity and inclusion, and ensuring employee well-being, AI empowers HR professionals to become more strategic partners within their organizations. In the years to come, the integration of AI and HR promises to further elevate the employee experience, ultimately benefiting both individuals and organizations alike.

Games

Overview

AI-driven games represent an exciting frontier in the world of interactive entertainment. By harnessing the power of artificial intelligence, these games offer players experiences that are not only immersive but also adaptive and dynamic. AI plays a pivotal role in shaping the gameplay, whether by creating intelligent non-playable characters, generating procedural content, or enabling personalized gaming experiences. It also extends its influence to game development, providing tools for content generation, quality assurance, and even assisting in the creation of interactive narratives. In essence, AI-driven games herald a new era where players can enjoy ever-evolving and uniquely tailored gaming experiences that continue to push the boundaries of innovation and immersion.

AI in Games

In the world of video games, AI is a transformative force, impacting every facet of the gaming experience. While the most prominent use of AI in games is seen in non-player character behavior, its influence extends to numerous other dimensions of gameplay, interaction, and immersion. NPC behavior, powered by AI, is the lifeblood of in-game worlds. These AI-controlled characters are far from mere automatons, as they exhibit intelligent reactions to a player's actions, follow scripted paths, engage in combat scenarios, and, crucially, make dynamic decisions that mold the game's storyline. This multifaceted approach to NPC behavior ensures that the game world feels alive and responsive, offering players a sense of autonomy and unpredictability.

Another pivotal aspect of AI in gaming is enemy AI, which enhances the excitement and thrill of gaming. AI controls the behavior of these virtual adversaries, allowing them to react intelligently to the player's tactics, creating challenging and adaptable combat encounters. In addition to behavior, AI powers pathfinding algorithms. These algorithms enable characters and units in games to navigate complex environments realistically, ensuring that NPCs and player-controlled characters move organically and avoid obstacles. This, in turn, contributes to a more immersive and realistic gaming experience.

AI's influence extends beyond behavior into procedural content generation. AI generates game content such as levels, maps, and terrain procedurally, introducing infinite variations and keeping the game world consistently fresh, preventing repetitive gameplay. Interactive story-driven games benefit from AI-driven NLP. This feature enables natural dialogues and interactions between players and characters, allowing players to engage in meaningful conversations with AI-driven NPCs.

Adaptive difficulty is yet another facet of AI, ensuring that the game experience remains enjoyable for players of all skill levels. AI dynamically adjusts the game's difficulty based on a player's skill and performance, maintaining a balance between challenge and accessibility.

AI-driven behavior tree systems model complex decision-making for characters. These trees determine how characters respond to stimuli, events, and objectives, rendering in-game interactions dynamic and unpredictable. AI excels in player profiling, analyzing behavior and preferences to personalize the gaming experience. AI can recommend in-game items, suggest challenges, or match players with similar skill levels, fostering a sense of connection and fair competition.

Voice and speech recognition, powered by AI, enhance interactivity. This technology allows players to issue voice commands and engage in voice interactions within games, providing a novel way to control in-game actions and enhance immersion.

AI's reach extends to character animations, making them appear more lifelike and emotionally expressive during cutscenes and interactions, enriching the narrative experience. Realistic physics simulation, driven by AI-powered physics engines, contributes to the gaming experience by simulating real-world physics, enabling realistic object interactions, destructible environments, and lifelike animations.

Dynamic game worlds are shaped by AI systems that modify the game environment in response to player actions. This includes spawning new enemies, triggering events, and altering the game environment based on player progress, ensuring a dynamic and engaging game world. AI significantly contributes to game testing and quality assurance, identifying bugs, glitches, and performance issues, leading to a smoother and more polished gaming experience for players.

Content moderation in online multiplayer games benefits from AI-powered tools that detect and prevent toxic behavior, cheating, and inappropriate content, creating a safer and more enjoyable gaming environment. AI aids game designers by providing assistance in generating level layouts, suggesting game mechanics, offering creative ideas for game development, streamlining the design process, and fostering innovation.

In summary, AI's multifaceted role in gaming, from NPC behavior and enemy AI to pathfinding, procedural content generation, and interactive story-driven elements, enhances the gaming experience and fosters a dynamic and evolving world of video games. As AI technology continues to advance, the future of gaming promises even more exciting prospects, offering increasingly sophisticated and dynamic gameplay experiences.

Sports

Overview

Artificial intelligence has made remarkable strides in the realm of sports, impacting athlete performance, fan engagement, and operational efficiency. It plays a pivotal role in the sports world, providing a competitive advantage to athletes, teams, and organizations while also elevating the overall fan experience.

The FIFA 2022 World Cup in Qatar harnessed the power of AI across various tournament aspects to improve the fan experience, optimize player performance, and enhance operational efficiency. Qatar established a technology hub that utilized AI to monitor spectators, predict crowd surges, and regulate stadium temperatures.

In the world of sports, renowned for its precision, strategy, and human potential, the integration of AI is rapidly reshaping the landscape. From performance analysis to fan engagement, AI's influence transcends traditional boundaries, ushering in a new era of data-driven decision-making, advanced training methodologies, and enriched fan experiences.

AI in Sports

AI has made a profound impact on sports in various dimensions. Performance analysis is one area where AI shines. AI-driven video analysis tools dissect game and practice footage to offer valuable insights into player performance, strategy, and opponent tendencies. Coaches and analysts leverage this data for informed, data-driven decisions, which lead to peak team performance. AI coaches complement human expertise by analyzing player and team data, offering alternative strategies, and highlighting areas for improvement, resulting in more comprehensive and effective coaching.

Player biometrics elevate athlete performance monitoring, with wearable devices equipped with AI capturing real-time data on vital signs, movements, and performance metrics. AI also plays a critical role in injury prevention and rehabilitation, predicting risks based on player biometric data and movements. These models then tailor rehabilitation programs to individual athletes, ensuring faster and more effective recovery.

In recruitment and scouting, AI leverages vast player statistics and performance metrics datasets to help teams identify promising talent. This ability to make informed decisions during player drafts and transfers is invaluable. Additionally, AI algorithms benefit fantasy sports enthusiasts by offering player recommendations, analyzing data, and suggesting lineup changes based on performance and matchups, enhancing the enjoyment of fantasy sports.

AI enhances fan engagement by providing personalized content recommendations, interactive experiences, and real-time statistics via mobile apps and websites. This makes the fan experience more immersive. AI-powered systems revolutionize the creation of video highlights and instant replays, engaging viewers more effectively. AI also supports referees and umpires by offering instant video reviews and decision support, reducing human errors and enhancing the accuracy and fairness of sports events.

On the business side of sports, AI plays a pivotal role in ticket pricing and sales optimization. AI models analyze data and market dynamics to optimize pricing, ensuring stadiums are filled while maximizing revenue, thereby contributing to the financial sustainability of sports. Broadcasts have significantly improved with AI by introducing virtual graphics, augmented reality elements, and player tracking data, delivering a superior viewing experience to fans. Even stadium operations and maintenance have benefited from AI-driven predictive maintenance systems, which monitor equipment and facilities to prevent breakdowns and ensure a smooth fan experience.

Athlete tracking relies on AI-driven computer vision systems capturing player movements and positions during games, providing real-time data for analysis and fan engagement. Also, performance wearables optimize the athlete's journey by tracking movements, muscle fatigue, and hydration levels, offering essential data for performance enhancement, injury prevention, and overall well-being.

In summary, AI's profound influence on sports is undeniable. AI's contributions are far-reaching, from enhancing performance and training regimens to improving the fan experience and optimizing the business side of sports. As technology continues to evolve, the future of sports holds the promise of even more advanced data-driven strategies, innovative fan engagement, and a heightened standard of athlete well-being and performance, underscoring the potential of AI to reshape sports as we know it.

Telco Analytics

Overview

AI plays a vital role in the telecommunications (telco) industry by improving network management, enhancing customer experiences, and optimizing operations. AI analytics can be used in many telecommunication applications such as network reliability, monitoring of cell phone towers, predicting churn, customer analysis, revenue and cost forecasting, and product analysis.

AI-driven telco analytics not only enhances network efficiency but also improves customer satisfaction, reduces operational costs, and enables telcos to remain competitive in a rapidly evolving industry. As AI technology advances, telcos will continue to find new ways to leverage AI for better service delivery and business outcomes.

The telecommunications industry is at the forefront of a technological revolution, with AI playing a pivotal role in transforming operations, enhancing user experiences, and ensuring optimal service quality. From network optimization to customer retention, AI is redefining how telcos operate, deliver services, and compete in a rapidly evolving digital landscape.

AI in Telco Analytics

The AI-driven transformation in telecommunications centers on network optimization, where algorithms meticulously analyze network performance data, including traffic patterns, congestion, and equipment status. This enables AI to predict and proactively address network issues, reducing downtime and ensuring optimal quality of service. AI also plays a pivotal role in resource allocation by optimizing bandwidth and spectrum allocation to manage user demand efficiently while minimizing operational costs.

Predictive maintenance is a crucial AI application that reinforces network stability by proactively identifying infrastructure issues and potential failures. AI enables scheduled maintenance to prevent service disruptions, maintaining the paramount quality of service (QoS) in telecommunications. Simultaneously, AI plays a pivotal role in continuously monitoring and improving service quality, ensuring the fulfillment of service level agreements (SLAs). This is achieved through the monitoring of service quality metrics and customer feedback.

Efficiency is paramount in customer service, and AI-driven chatbots represent a beacon of efficiency. These chatbots instantly respond to customer queries, troubleshoot issues, and assist with account management, reducing response times and enhancing overall efficiency. AI's predictive analytics is a game changer for understanding customer behavior, predicting preferences, and equipping telcos with insights to minimize customer churn and foster loyalty by customizing marketing, products, and pricing.

AI plays an indispensable role in revenue assurance by diligently analyzing billing and revenue data, ensuring accurate billing and robust revenue collection, and safeguarding the financial health of telcos. Market analysis benefits from AI's ability to synthesize and analyze vast amounts of data, assisting telcos in defining their business strategy and product development to remain competitive and responsive to market dynamics.

The growth of IoT and machine-to-machine (M2M) connectivity presents new challenges and opportunities for telcos, which AI efficiently manages to ensure optimal data transmission and network performance. Voice and speech recognition powered by AI redefine customer interactions, facilitating communication through voice commands, automated phone systems, and virtual assistants, enhancing self-service options and streamlining interactions.

Fraud detection and prevention are critical in the telecommunications industry, with AI analyzing call detail records and usage patterns to identify fraudulent activities such as SIM card cloning and call spoofing, preserving the integrity of the network and protecting users. AI also serves as a guardian, with AI-powered security solutions instrumental in detecting and responding to cyber threats, such as malware, distributed denial-of-service (DDoS) attacks, and intrusions. These systems analyze network traffic patterns to identify unusual behavior and shield telcos against vulnerabilities, ensuring network and user security.

In summary, AI's role in telecommunications is transformative. From optimizing network performance and maintaining network infrastructure to enhancing customer support and guarding against cybersecurity threats, AI touches all aspects of the telco landscape. With AI as their ally, telecommunications companies are better positioned to meet the demands of an increasingly connected world, offer higher-quality services, and foster lasting customer relationships.